The Electrician's Troubleshooting and Testing Pocket Guide

The Electrician's Troubleshooting and Testing Pocket Guide

John E. Traister
H. Brooke Stauffer

McGraw-Hill

New York San Francisco Washington, D.C. Auckland Bogotá
Caracas Lisbon London Madrid Mexico City Milan Montreal
New Delhi San Juan Singapore Sydney Tokyo Toronto

Library of Congress Cataloging-in-Publication Data

Traister, John E.
   The electrician's troubleshooting and testing pocket guide / John E. Traister, Brooke Stauffer.—Rev. ed.
      p. cm.
   Rev. ed. of: The electrician's testing and troubleshooting pocketguide / John E. Traister. 1996.
   ISBN 0-07-135472-7
   1. Electric testing. 2. Electric apparatus and appliances—Maintenance and repair. 3. Electric motors—Maintenance and repair. 4. Electriccircuit analysis. I. Stauffer, Brooke. II. Traister, John E. Electrician's testing and troubleshooting pocket guide. III. Title.
TK401.T695 2000
621.37—dc21

                                                                                                          99-089959
                                                                                                            CIP

McGraw-Hill

A Division of The McGraw·Hill Companies

Copyright © 2000 by The McGraw-Hill Companies, Inc.. All rights reserved. Printed in Canada. Except as permitted under the United States Copyright Act of 1976, no part of this publication may be reproduced or distributed in any form or by any means, or stored in a data base or retrieval system, without the prior written permission of the publisher.

      5 6 7 8 9 0 WEB/WEB 0 9 8 7 6 5 4 3

ISBN 0-07-135472-7

The sponsoring editor for this book was Zoe G. Foundotos, the editing supervisor was Sally Glover, and the production supervisor was Pamela Pelton. It was set in the JJB1 design in 1 Stone Serif by Deirdre Sheean, Paul Scozzari, Kim Sheran, and Michele Pridmore of McGraw-Hill's Professional Book Group composition unit, Hightstown, New Jersey.

Printed by Webcom

 This book was printed on acid-free paper.

McGraw-Hill books are available at special quantity discounts to use as premiums and sales promotions, or for use in corporate training programs. For more information, please write to the Director of Special Sales, Professional Publishing McGraw-Hill, Two Penn Plaza, New York, NY 10121-2298. Or contact your local bookstore.

Information contained in this work has been obtained by The McGraw-Hill Companies, Inc. ("McGraw-Hill") from sources believed to be reliable. However, neither McGraw-Hill nor its authors guarantee the accuracy or completeness of any information published herein, and neither McGraw-Hill nor its authors shall be responsible for any errors, omissions, or damages arising out of use of this information. This work is published with the understanding that McGraw-Hill and its authors are supplying information but are not attempting to render engineering or other professional services. If such services are required, the assistance of an appropriate professional should be sought.

CONTENTS

Introduction iii

1 Analog Meters *1*
2 Miscellaneous Testing Instruments *27*
3 Digital Multimeters *37*
4 Basics of Troubleshooting *51*
5 Troubleshooting Dry-Type Transformers *57*
6 Troubleshooting Fluorescent Lamps and Fixtures *65*
7 Troubleshooting Incandescent Lamps and Fixtures *83*
8 Troubleshooting HID Lamps and Fixtures *89*
9 Troubleshooting Electric Motors *103*
10 Troubleshooting Motor Bearings *193*
11 Troubleshooting Contactors and Relays *211*

12 Troubleshooting Power Quality Problems *227*

Appendix: Electrical Fundamentals *241*

Index 283

Introduction

Electrical measuring and testing instruments are crucial in the installation, troubleshooting, and maintenance of electrical systems of all types, particularly in commercial and industrial facilities. Electricians and technicians involved with installing, maintaining, and repairing electrical equipment need a good working knowledge of portable testing instruments and how they can best be used to diagnose and fix problems in the field.

This new, completely updated edition of *The Electrician's Troubleshooting and Testing Pocket Guide* is an invaluable resource for electricians and technicians using portable meters to test, maintain, and repair all types of electrical equipment and systems.

Most operational problems of electrical equipment and systems involve one of four basic faults:

- Short circuit
- Ground fault
- Open circuit
- Change in electrical value

This guide describes troubleshooting techniques to identify such problems using portable field-testing instruments. Although it covers many types of test equipment, this book emphasizes the use of digital multimeters (DMMs), the most common and versatile electrician's diagnostic tool in use today.

This revised edition also includes a new chapter on troubleshooting power quality problems. The rapidly increasing use, in commercial and industrial facilities, of computers and other solid-state electronic equipment such as dimming fluorescent and HID ballasts is affecting the operation and safety of electrical distribution systems. In 1996, the National Electrical Code was revised to acknowledge the seriousness of this concern, and over the last few years diagnosing and solving power quality problems has become a major priority for many electricians and maintenance personnel.

Scope of This Book

The Electrician's Troubleshooting and Testing Pocket Guide covers the use of DMMs and other testing equipment to troubleshoot electrical and electronic circuits used for power and control applications. In general, it concentrates on traditional electromechanical and inductive equipment found in commercial and industrial occupancies—equipment such as motors, transformers, and lighting. In general, this guide does not cover testing and troubleshooting of the following types of equipment and systems:

Communications systems. The use of network cable analyzers, optical time domain reflectometers (OTDRs), optical power meters, and other equipment used for testing and troubleshooting communications systems such as telecommunications, computer local area networks (LANs), and outside plant fiber-optic installations are outside the scope of this publication.

Electronic components and systems. This book touches on testing of electronic components such as resistors, small capacitors, and diodes. However, the broad subject of troubleshooting electronic components and circuits using digital multimeters and other portable test equipment is covered in much greater detail in a different McGraw-Hill publication:

Electronic Troubleshooting and Repair Handbook
McGraw-Hill, Inc.
By Homer L. Davidson
1995
ISBN 0-07-015676-X (H)

Brooke Stauffer
Director, Codes and Standards
National Electrical Contractors Association (NECA)
Bethesda, Maryland

CHAPTER 1

Analog Meters

Traditional meters used by electricians and technicians for field testing and troubleshooting are analog type. In an analog meter, the magnitude of the property being measured (voltage, current, resistance, illuminance, etc.) is translated into a corresponding physical movement of a pointer, needle, or other indicator. Higher voltage, for example, is shown by the needle of a traditional voltmeter swinging farther to point at a higher number on a dial.

Analog meters are generally limited to a single function. Thus we have ammeters, voltmeters, and resistance testers (frequently called *meggers* in the field, after the name of one of the best-known brands of resistance tester). In some cases the usefulness of traditional analog electrical test instruments can be extended or modified with special adaptors or sensors; some voltmeters, for example, can also be used to measure temperature.

Over the last decade, the different types of single-function analog meters have been largely replaced by digital (computerized) meters that combine many measurement functions within a single compact unit. These digital multimeters, or DMMs, are now used for most testing, troubleshooting, and maintenance purposes. However, there are still many older analog meters in use, and a general knowledge of their properties is useful to electricians and technicians.

This chapter briefly describes the various types of analog electrical meters and instruments, and how they are used. The rest of the book, starting with Chapter 3, concentrates primarily on the use of DMMs.

Ammeters

Figure 1-1 shows a clamp-on ammeter used to measure current in an electrical circuit while the circuit carries the full load. Although the exact operating

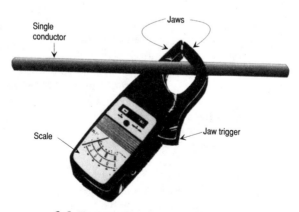

1-1 Typical clamp-on type ammeter.

procedures vary with the manufacturer, most operate as follows when measuring current:

Step 1. Make sure the battery-attachment case (for use when in the ohmmeter mode) is removed from the instrument.

Step 2. Release on pointer locks.

Step 3. Turn the scale selector knob until the highest current range appears in the scale window.

Step 4. Press the trigger button to open the jaws of the clamp before encircling one of the conductors under test with the transformer jaws.

Caution

Never encircle two or more conductors; only encircle one conductor as shown in Figure 1-1.

Step 5. Release finger pressure on the trigger slowly to allow the jaws to close around the conductor and keep an eye on the scale while doing so. If the pointer jumps abruptly to the upper range of the scale before the jaws are completely closed, the current is probably too high for the scale used. Should this happen, remove the jaws immediately from the conductor and use either a higher scale or a range-extender attachment as discussed in the next section. If the pointer deflects normally, close the jaws completely and take the reading from the scale.

Extending the Range of Ammeters

The range of an ac ammeter can be extended by using a range extender as shown in Figure 1-2. This device permits taking a measurement of higher current beyond the range of the regular clamp-on ammeter. The model shown extends the current range ten times to allow an actual current reading of 1000 A on a 0- to 100-A meter scale. To illustrate its use, if the scale shows a reading of, say, 42 A, the actual current (using the range extender) would be 420 A because $42 \times 10 = 420$.

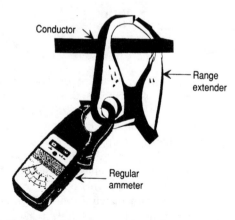

1-2 Clamp-on transformer extends ammeter range.

Current Multipliers

Sometimes it is desirable to use a current multiplier with a clamp-on ammeter, such as the one shown in Figure 1-3. This device allows current measurement on low-current equipment since the load current shown has been multiplied either two, five, or ten times; that is, if the meter scale shows a reading of, say, 62 A and the 10× multiplier was used, the actual load current would be:

$$\frac{62}{10} = 6.2 \text{ A}$$

Precautions

When using clamp-on ammeters, care must be taken to obtain accurate readings. Some items to be considered include:

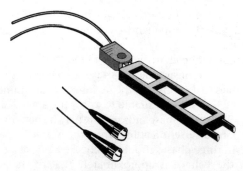

1-3 Current multiplier in use.

1. Make certain that the frequency of the circuit under test is within the range of the instrument. Most are calibrated at 70 hertz (Hz).

2. Take care that stray magnetic fields do not affect the current reading; that is, arrange the meter leads as far away as possible from the conductors under test.

Try to take current readings in a control panel at a location remote from magnetic relays that might influence the accuracy of the reading.

Avoid taking current readings on conductors at a point close to a transformer.

3. When current readings are taken on high-voltage conductors, always use a hot-line extension pole specifically designed for use with a high-voltage clamp-on ammeter.

Ammeter Applications

The ammeter is quite handy for troubleshooting various electrical components by indicating a change in electrical value. Many examples, including numerous troubleshooting charts, may be found throughout this book. In the meantime, let's take a couple of simple examples to demonstrate the usefulness of an ammeter.

Sometimes it is desirable to know the approximate load on a three-phase motor. This may be determined by taking a current reading while the motor is operating. All three phases should be checked, one at a time.

If the voltage is approximately that of the rated voltage of the motor, and the ammeter shows that the motor is drawing a current close to the nameplate rating, it may be reasonably assumed that the motor is carrying a full load. If the ampere reading is much less, the motor is not carrying a full load. If the current is much higher than the nameplate rating after the motor has come up to full speed, it may be assumed that the motor is overloaded.

To determine if, say, an electric baseboard heater is operating properly, examine the nameplate to find the heater's characteristics. Let's assume that the nameplate indicates a 1000-W single-phase, two-wire heating element operating at 240 A. If an ammeter reading is taken while the heater is operating and the reading shows approximately 4 A of current, it may be assumed that the heater is working as it should, because:

$$I = \frac{P}{E} \text{ or } \frac{1000}{240} = 4.16 \text{ A}$$

If the ampere reading is much different from 4A, some fault in either the circuit or the heater is present.

Recording Ammeters

When it is desired to have a continuous and/or permanent record of the current in a given electrical circuit, a graphic or recording instrument may be used. Such an instrument has a meter element similar to the conventional indicating ammeters, but, in addition, it is equipped with a pen or other marking device so that a curve is drawn as current changes occur. The marking device on the recording instrument replaces

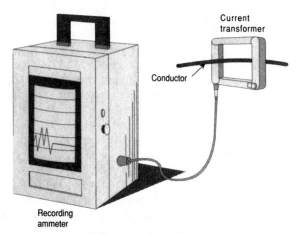

1-4 Recording ammeter.

the pointer on scale-indicating meters, and the marking device traces a line on a chart representing the value of the amperes that the instrument is measuring (see Figure 1-4).

The charts are usually either circular or in strip form, depending on the type of instrument.

Voltmeters

The unit of electromotive force (emf) is the volt (V). One volt is a form of pressure that, if steadily applied to an electrical circuit having a resistance of 1 Ω, produces a current of 1 A.

Voltmeters are used to accurately measure the pressure or voltage in various electrical circuits. Very low voltage values are measured in millivolts (1 volt = 1000 mV) by a millivoltmeter with low resistance.

A voltmeter should be connected across the terminals at the place the voltage is to be measured, as shown in Figure 1-5. A voltmeter should never be connected across a circuit having a voltage higher than the rating of the instrument; that precaution must be observed particularly in the case of measurements with a millivoltmeter.

When connecting a voltmeter to a dc circuit, always observe the proper polarity. The negative lead of the meter must be connected to the negative terminal of the dc source, and the positive lead to the positive terminal.

If the leads are connected to opposite terminals, the needle will move in the reverse direction. Since the voltage constantly reverses polarity in an ac circuit, there is no need to observe polarity when measuring voltage on ac circuits (see Figure 1-6).

Many portable voltmeters are designed with two or more voltage ranges that can be read on a common scale, such as 0 to 150 V, 0 to 300 V, and 0 to 600 V.

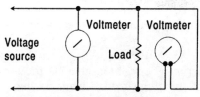

1-5 Voltmeter connection methods.

1-6 Checking voltage at a 125-V duplex receptacle.

(See Figure 1-7.) When using a multirange voltmeter, it is best to select a higher range than needed to assure that no damage will occur to the instrument. Then, if the initial reading indicates that a lower scale is needed to obtain a more accurate reading, the meter can be switched or otherwise adjusted to the next lowest range.

One of the reasons for using various ranges of voltmeters is that the greatest accuracy is obtained on the upper half of the scale. Therefore, if a single 0- to 600-V range were used, lower voltages would be harder to read and meter accuracy would be less.

Voltmeter Applications

Voltmeters can be used for other electrical tests, such as troubleshooting circuits, circuit tracing, and measuring low resistance. For example, a common cause

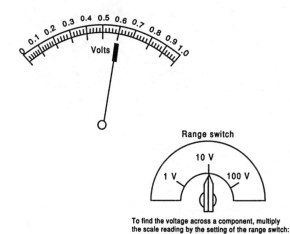

1-7 Multirange, one-scale voltmeter.

of electrical problems is low equipment voltage. This problem usually occurs because of one or more of the following reasons:

- Undersized conductors
- Overloaded circuits
- Transformer taps set too low

Low-Voltage Test

When making a low-voltage test, first take a reading at the main switch or service entrance. If, for example, the main service is 120/240, single-phase, three-wire, the voltage reading between phases (ungrounded conductors) should be between 230 and 240 V. If the reading is much lower than this, the fault lies with the utility company supplying the power, and they should be notified to correct the problem. However, if the reading at the main service switch is between 230 and 240 V, the next procedure is to check the voltage reading at various outlets throughout the system.

When a low-voltage problem is found on a circuit, leave the voltmeter terminals connected across the line and begin disconnecting all loads, one at a time, that are connected to the circuit. If the problem is corrected after several of the loads have been disconnected, the circuit is probably overloaded. Steps should be taken to reduce the load on the circuits or else increase the wire size to accommodate the load.

As mentioned previously, loose connections can also cause low voltage; to check for this, the entire circuit should be de-energized, and each terminal in all disconnect switches, motor starters, and so on, should be checked for loose connections. A charred or blackened terminal screw is one sign to look for in the various components of the system.

Ground Fault

Another problem is a ground fault in a circuit. For example, assume that a three-phase, three-wire, 240-V, delta-connected service is used for power in a small industrial plant. The service equipment is installed as shown in Figure 1-8. Under proper operating conditions, the voltmeter should read 240 V between phases

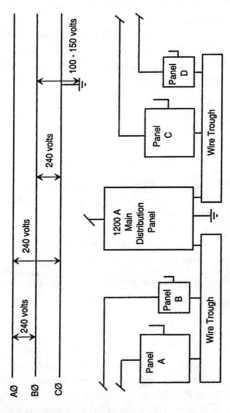

1-8 Diagram of a small industrial electric service.

(A and B, B and C, A and C), and approximately 150 V between each phase to ground.

However, if on checking with the voltmeter, two of the readings show a voltage of 230 V between phase and ground and the other reading is only 50 V between phase and ground, then the phase with the lowest reading (50 V) has a partial ground or ground fault.

The sequence for correcting the ground fault is as follows:

Step 1. Connect one of the voltmeter terminals to the grounded main-switch housing and the other to the phase terminal that indicated the ground fault.

Step 2. Disconnect switch A and check the voltmeter reading. If no change is indicated, disconnect switch B, switch C, and so on, until the voltmeter shows a change, that is, a reading of approximately 150 V from phase to ground.

Step 3. Assuming that the voltmeter indicates this reading when switch D is thrown to the OFF position, we then know that the ground fault is located somewhere on this circuit.

Step 4. Switch D disconnects the 400-A circuit feeding eight 15-horsepower (hp) motors and connected as shown in Figure 1-9. One voltmeter lead is connected to the grounded housing of switch D and the other lead to one of the phase terminals. The switch is then thrown to the ON position. Check each phase terminal until the one with the ground fault is located.

Step 5. Then the motors, one at a time, are disconnected from the circuit until the one causing the trouble is found; that is, when the motor or motor circuit having the ground fault is disconnected, the voltmeter will show a normal voltage of approximately 150 V from phase to ground.

Step 6. The faulty motor or motor circuit can then be corrected according to standard maintenance procedures.

When testing electrical circuits with a voltmeter, it is usually best to begin at the main service disconnect. First, test the voltage on the line side to see if the line

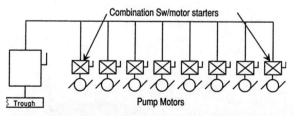

1-9 Wiring diagram for eight 15-hp pump motors fed from a 400-A safety switch.

is "hot" from the outside service wires; if it is, then test the fuses. The fuses may be checked by testing across diagonally from the line to the load side as shown in Figure 1-10.

The voltmeter is a very useful instrument for all personnel working with electricity, but its capabilities must be thoroughly understood in order to obtain its full usefulness. Proper care and maintenance are also necessary to assure accurate readings and reliability. Figure 1-11 shows two popular kinds of voltmeters. Meter A is a combination volt-ohm-ammeter with a conventional swinging pointer to indicate the reading; meter B works on the "ting" principle—similar to an air gauge—and gives only approximate voltage readings.

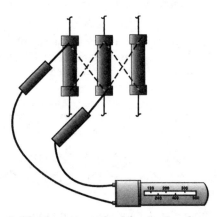

1-10 Correct method for testing fuses with a voltmeter.

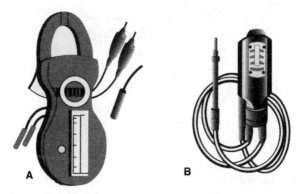

1-11 Some common types of voltmeters.

Megohmmeters

A typical megohmmeter (megger) is composed of a hand-driven or motor-powered ac generator and/or a transformer with voltage rectified to 100, 250, 500, and 1000 V dc, a cross-coil movement with 0 to 20,000-ohm (Ω) and 0- to 1000-megohm (MΩ) scales, a carrying case, and test leads. The megger (Figure 1-12) is used to measure the resistance in megohms to the flow of current through and/or over the surface of electrical equipment insulation. The test results are used to detect the presence of dirt, moisture, and insulation deterioration. The instrument also typically measures resistance up to 20,000 W.

The instruction manuals accompanying the megohmmeter contain detailed instructions about

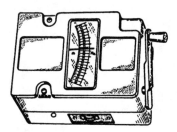

1-12 Typical hand-driven megohmmeter.

preparing for tests and connecting the unit to various types of equipment. Figures 1-13 and 1-14 give some practical points on hookups of megger instruments.

Figure 1-13 shows that ac motors and starting equipment can be tested by connecting one side of the megger to the motor side of the main switch and the second test connection to a clip on the motor housing. If an insulation weakness is indicated by this test, the motor and starter should be checked separately.

Figure 1-14 shows the connection for testing low-voltage power cable. After both ends of the cable have been disconnected, the conductors are tested, one at a time, by connecting one of the leads to the conductor under test and connecting the remaining conductors (within the cable) to ground and then to the other (ground) test lead. Similarly, other insulation resistance, such as between conductor and outside protective sheath and between conductors, can be measured.

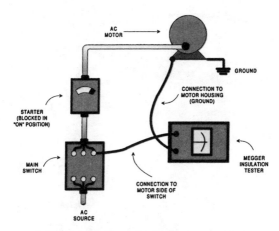

1-13 Connections for testing ac motors and starting equipment.

Lighting and distribution transformers are tested by first making sure that switches or circuit breakers on both the primary and secondary sides are open. The high-voltage winding to ground and low-voltage winding to ground are consecutively checked by separate tests. The resistance between the two is then checked with neither of them grounded.

Testing dc Motors and Generators

To test dc motors and generators, disconnect the apparatus from any power source or load and attach the negative test lead of the megohmmeter to the machine ground and the positive lead to the brush

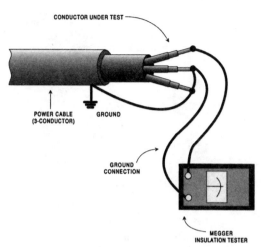

1-14 Megger connection for testing low-voltage power cable.

rigging. Measuring the insulation resistance in this manner indicates the overall resistance of all components of the unit.

To measure the insulation resistance of the field or armature alone, either remove the brushes or lift them free of the commutator ring and support them by means of a suitable insulator. Connect one test lead to the frame ground and the other to one of the brushes. Insulation resistance of the field alone will then be indicated as shown in Figure 1-15. With the brushes still removed from the commutator ring, connect one of the megger test leads to one of the segments of the commutator and the other to the frame ground. The insulating resistance of the armature alone will then be indicated. This test may be repeated for all segments of the commutator.

Testing ac Motors

To test ac motors, first disconnect the motor from the line, either by use of the switch or by disconnecting the wiring at the motor terminals. If the switch is used, remember that the insulation resistances of the connecting wire, switch panel, and contacts will all be measured at the same time. Connect the positive megger lead to one of the motor lines and the negative test lead to the frame of the motor, as shown in

1-15 Megger connections for testing dc motors and generators.

Figure 1-16. If insulation resistance minimums have been established, the reading can be checked against them.

Testing Circuit Breakers

Circuit breakers may be tested with the megger by first disconnecting the circuit breaker from the line and connecting the megger black lead to the frame or ground. Check the insulation resistance of each terminal to ground by connecting the red (positive) lead to each terminal in turn and making the measurements. Next open the breaker and measure the insulation resistance between terminals by putting one lead on one terminal and the other on the second for a two-terminal breaker; for a three-pole breaker, check between poles 1 and 2, 2 and 3, and 1 and 3, in turn.

Testing Safety Switches and Switchgear

Switches with safe insulation are a vital part of any electrical installation and so deserve careful attention. Switches should be completely disconnected from the line and relay wiring before testing. When manual switches are being tested, measure the insulation resistance from ground to terminals and between terminals. When testing electrically operated switches, check the insulation resistance of the coil or coils and contacts. For coils, connect one megger lead to one of the coil leads and the other to ground. Next, test between the coil lead and core iron or solenoid element.

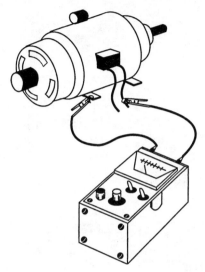

1-16 Method of testing an ac motor.

If relays are involved, measure insulation resistance with one lead connected to the relay plate or contact and the other test lead connected to the coil, core, or solenoid contact.

Ground Resistance

Figure 1-17 shows the simplest method for testing the resistance of earth. The direct or two-terminal test consists of connecting terminals P_1 and C_1 of the megohmmeter to the ground under test, and terminals P_2 and C_2 to an all-metallic water-pipe

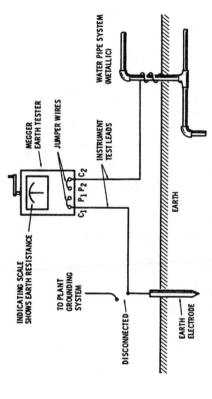

1-17 Direct method of earth-resistance testing.

system. If the water-pipe system covers a large area, its resistance should only be a fraction of an ohm and, therefore, the meter reading will be that of the ground or electrode under test.

CHAPTER 2

Miscellaneous Testing Instruments

Ammeters, voltmeters, and megohmmeters are the most common traditional analog devices used for field-testing and troubleshooting applications. However, there are a number of more specialized types of test instruments that also should be mentioned briefly.

Frequency Meter

Frequency is the number of cycles completed each second by a given ac voltage, usually expressed in hertz; 1 hertz = 1 cycle per second.

The frequency meter is used in ac power-producing devices like generators to ensure that the correct frequency is being produced. Failure to produce the correct frequency will result in heat and component damage.

There are two common types of frequency meters. One operates with a set of reeds having natural vibration frequencies that respond in the range being tested. The reed with a natural frequency closest to that of the current being tested will vibrate most strongly when the meter operates. The frequency is read from a calibrated scale.

A moving-disk frequency meter works with two coils, one of which is a magnetizing coil whose current varies inversely with the frequency. A disk with a pointer mounted between the coils turns in the direction determined by the stronger coil. Solid-state frequency meters are also available.

Power-Factor Meter

Power factor is the ratio of the true power to the apparent power, and it depends on the phase difference between the current and the voltage. The single-phase power-factor meter in Figure 2-1 is so constructed that the rotating field is produced by the line voltage, and at a power factor of 100 percent, or unity, there is no torque on the moving coil and the pointer rests at the center of the scale, or 100. Depending on whether the current is leading or lagging when the power factor is less than unity, the pointer swings to the left or to the right of the scale, indicating directly the value of the power factor.

Three-phase power-factor meters are also available and are quite common on industrial switchboard installations. Since most industrial establishments are charged a penalty if the power factor falls below 90 percent, industries try to maintain a high power factor at all times. Two other reasons for trying to maintain a power factor near unity are:

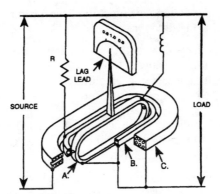

2-1 Power-factor meter.

- The reduction of reactive current provides more capacity for useful current on the mains, feeders, and subfeeders.
- A high power factor provides better voltage regulation and stability.

Synchroscopes

When two alternators are about to be connected in parallel, the voltages of the two must be approximately the same; their voltages must be exactly in phase; and their frequencies must be approximately the same. If these differences are too great, the alternators are likely to pull entirely out of phase, thus causing a complete shutdown.

Figure 2-2 shows a schematic drawing of the connections for a synchroscope in a high-voltage generating system. The synchroscope shown provides the following functions:

- It indicates whether the generator is running too slow or too fast.
- It indicates the amount by which the generator is slow or fast (the difference in frequency) at any instant.
- It indicates the exact time of coincidence in phase relationship between all generators connected to the system.

When the synchroscope shows that both alternators are synchronized, the switches may be closed to allow both alternators to work together in parallel. The ideal indication, on most synchroscopes, is when the pointer either stops or moves very slowly at mid-scale.

Tachometers

A tachometer is a device used to indicate or record the speed of a machine in revolutions per minute (rpm).

Most of the better hand-held models have two buttons: a stop button to hold readings and a second button to release the pointer to 0. The ball-bearing spindle on these devices is placed against a rotating object on the machine such as the motor drive shaft, and the speed of the shaft is read directly on the dial of the tachometer.

Totally enclosed rotating equipment may be checked for speed by using a vibrating-reed tachometer. These tachometers operate on the well-known and time-tested principle of resonance. The instrument is

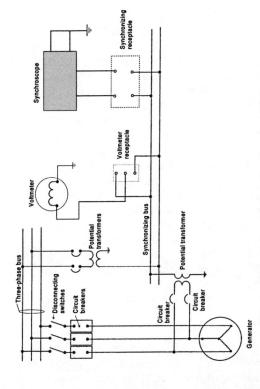

2-2 Synchronizing high-voltage generating system.

simply held against the motor, turbine, pump, compressor, or other rotating equipment and the speed is shown by the vibration of a steel reed which is tuned to a certain standard speed.

A photo tachometer uses a light that is aimed at the rotating shaft on which there is a contrasting color such as a mark, a chalk line, or a light-reflective strip or tape. The rotational speed in rpm is conveniently read directly from the indicating scale of the instrument. This tachometer design is especially useful on relatively inaccessible rotational equipment such as motors, fans, grinding wheels, and other similar machines where it is difficult, if not impossible, to make contact with the rotational unit.

An electrical tachometer consists of a small generator that is belted or geared to a unit whose speed is to be measured. The voltage produced in the generator varies directly with the speed of the rotating part of the generator. Since this speed is directly proportional to the speed of the machine under test, the amount of the generated voltage is a measure of the speed. The generator is electrically connected to an indicating or recording instrument which is calibrated to indicate units of speed such as rpm, fps, fpm, etc.

Footcandle Meter

A footcandle meter consists of light-barrier layer cells and a meter enclosed in a suitable covering that is capable of reading light intensities from 1 to 500 footcandles or more.

To use the footcandle meter, first remove the cover. Hold the meter in a position so the cell is facing toward the light source and at the level of the work plane where the illumination is required. The shadow of your body should not be allowed to fall on the cell during tests. A number of such tests at various points in a room or area will give the average illumination level in footcandles. Readings are taken directly from the meter scale.

Electrical Thermometers

For the measurement of temperatures, three basic electrical thermometer methods are used.

The resistance method makes use of the fact that the resistance of a metal varies in direct proportion to temperature. This method is normally used for temperatures up to approximately 1500°F.

The thermocouple method is based on the principle that a difference in temperature in different metals generates a voltage and is used for measuring temperatures up to about 3000°F.

The radiation pyrometer and optical pyrometer are generally used for temperatures above 3000°F. They combine the principle of the thermocouple with the effect of radiation of heat and light.

Phase-Sequence Indicator

A common phase-sequence indicator is designed for use in conjunction with any multimeter that can measure ac voltage. Most can be used on circuits with line voltages up to 550 V ac, provided the instrument used with the indicator has a rating this high.

To use the phase-sequence indicator, set the multimeter to the proper voltage range. This can be determined (if it is not known) by measuring the line voltage before connecting the phase-sequence indicator. Next, connect the two black leads of the indicator to the voltage test leads of the meter. Connect the red, yellow, and black adapter leads to the circuit in any order and check the meter for a voltage reading.

If the meter reading is higher than the original circuit voltage measured, then the phase sequence is black-yellow-red. If the meter reading is lower than the original circuit voltage measured, then the phase sequence is red-yellow-black. If the reading is the same as the first reading, then one phase is open.

Infrared Sensing Device

An infrared sensing device is an optical device that measures the infrared heat emitted from an object.

Cable-Length Meter

A cable-length meter measures the length and condition of a cable by sending a signal down the cable and then reading the signal that is reflected back. These instruments are also called time-domain reflectometers (TDRs).

Power Quality Analyzers

Power quality analyzers are portable test instruments similar in construction to the digital multimeters described in greater detail in Chapter 3. However, unlike DMMs, which typically measure only one property of electrical circuits at a time, power quality analyzers have dual probes that allow both voltage and current to be measured simultaneously. Power quality analyzers can also measure frequency and harmonics.

The results of these readings are displayed graphically, as shown in Figure 2-3. The ability to measure

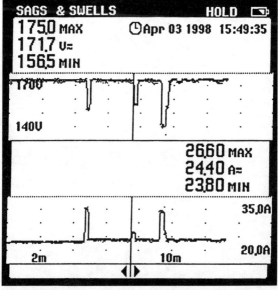

2-3 Power quality analyzer display showing voltage on top, current on bottom, time stamp at upper right.

and display multiple circuit characteristics at the same time is useful in troubleshooting power quality problems in power distribution systems. This subject is covered more fully in Chapter 12.

CHAPTER 3

Digital Multimeters

The five core functions of hand-held meters have always been measuring ac and dc voltage, ac and dc current, and resistance. Digital multimeters (DMMs) containing microprocessors still perform these same functions, but their built-in computing power allows them to offer other capabilities as well:

- Greater accuracy
- Better displays
- Accessory adapters for taking additional types of measurements
- Data-handling capabilities

Figure 3-1 shows a typical digital multimeter. The range of features, options, and accessories offered on DMMs varies widely from one brand and model to the next. The most important are described below.

Greater accuracy
The accuracy of DMM readings is typically from 0.5 to 0.1 percent, and results can be displayed to two or three decimal places. While this level of accuracy is not always needed for field troubleshooting of electromechanical

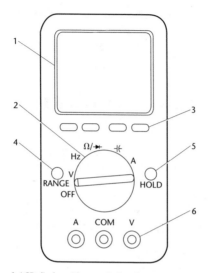

1 LCD display with numerical readout.
2 Measurement function knob.
3 Soft-keys – Use with measurement function knob to select measurements.
4 Range button – Use to set measurement range.
5 Hold button – Use to freeze display.
6 Input connectors

Note: Some DMMs have a separate function knob setting and/or input connector for A/mA.

3-1 Digital multimeter (DMM).

equipment, it can be useful in applications involving electronic circuits.

Better displays

Digital multimeter displays show numerals and graphical patterns (such as waveforms) rather than swinging needles. Displays are often large enough to read from a distance, and some are big enough to display two or more items simultaneously, such as voltage and frequency.

Most DMMs have a liquid-crystal diode display that expresses readings in contrasting shades of gray. Many models also have a backlighting switch for taking readings under poorly lighted areas. Maximum display readouts are always one digit less than the marked range. For example, the 200-ohm resistance range reads between 0.0 and 199.9Ω (Figure 3-3, later in this chapter). If higher resistance is present, "OL" or "1" (overlimit or out-of-range indication) shows in the display. When this happens, the rotary switch should be rotated to a higher range.

Hold, freeze, or capture mode

On many DMMs, pressing a "hold" button freezes a reading on the display screen so that the meter can be taken to a more convenient area for viewing. This feature is particularly useful in tight spaces with poor visibility, or when it isn't convenient to read the display at the same time you're taking a measurement on a circuit or piece of electrical equipment.

Construction/convenience features

Most digital multimeters have a shock-resistant heavy-duty case with a belt holster, and a tilt stand for placing on flat surfaces such as a table. Many also have handles that allow them to be hung at eye level, an advantage in many troubleshooting applications where space is tight. DMMs are very rugged and can last for years of trouble-free operation under heavy-duty use.

Many units can operate with the same 9-V battery for 2000 to 3000 hours because the solid-state circuits and LCD display have a very low current drain. Some models constantly display a battery status icon on the screen. In other models, a "Lo Bat" warning appears or the decimal point in the digital display blinks when the battery is nearing its end of life.

Function selection
DMMs have a dial or rotary switch that lets you select basic measurement functions (voltage, current, resistance, frequency, temperature, etc.). Higher-priced DMMs also have either four or eight "soft keys." These are pushbuttons whose function depends upon the type of measurement selected. When the dial is rotated to select a basic measurement function, such as current, some or all of these soft keys may become active. When this happens, the purpose of that key is displayed at the bottom of the LCD display (i.e., just above the soft keys). For some measurement functions, not all soft keys will be active.

Inputs and test leads
Most DMMs have three test jacks or inputs: voltage (V), current (A), and common or return (COM). The inputs marked V and A are normally colored red, as are the various test leads that plug into them. The common

input, which is used for all measurement functions, is normally colored black, as is the common test lead that plugs into it. *NOTE: Some units also have a fourth separate input for current measurements in the milliampere (mA) or microampere (μA) range.*

Accessories

DMM manufacturers offer a wide array of accessories that both extend measurement ranges and allow the instrument to be used for additional types of measurements, including:

- Power
- Power factor
- Energy (kWh)
- Harmonics
- Temperature (single probe, and dual probe for differential)
- Light intensity
- Relative humidity
- Carbon monoxide (CO)
- Airflow

General Instructions for Using Digital Multimeters

Because exact capabilities and features of different DMMs vary, it is important to read the manufacturer's manual supplied with the unit. The following procedures apply to digital multimeters generally.

Measuring voltage

Select a voltage measurement range. Connect test leads to the V and COM inputs. Place the DMM in parallel with the voltage source and load to measure voltage (Figure 3-2). Never place the meter in series with the circuit when measuring voltage.

Measuring current

Select a current measurement range. Connect test leads to the A and COM inputs. Place the DMM in series with the voltage source and load to measure current. Never place the meter across (in parallel with) the circuit when measuring amperes. The current in solid-state circuits such as printed circuit boards is measured in milliamperes (mA) or microamperes (µA) (Figure 3-3).

Measuring resistance

Select resistance test (Ω). Plug the red test lead into the voltage (V) input and the black lead into the common (COM) input. Place the probe tips across the suspected resistor or leaky component. A good resistor should read within plus or minus 10 percent of its rating. Thus, a sound 330-Ω resistor would register between 300 and 360 Ω (suspect a burned resistor is the reading is less than 300 Ω). It may be necessary to isolate the resistor or other component from the circuit to get an accurate reading (Figure 3-4).

Testing continuity

Select resistance test (Ω). Connect test leads to the V and COM inputs. Some DMMs sound a constant tone or noise when making continuity and diode tests. A constant tone indicates proper continuity. No tone (or a broken, stop-start sound) indicates an open circuit, intermittent faults, or loose connections (Figure 3-5).

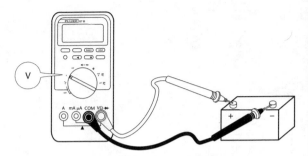

3-2 Measuring voltage.

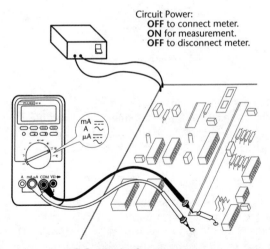

Circuit Power:
OFF to connect meter.
ON for measurement.
OFF to disconnect meter.

3-3 Measuring current.

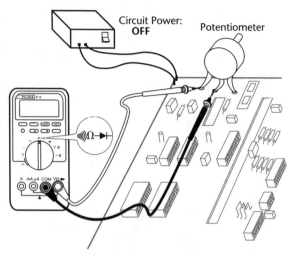

3-4 Measuring resistance.

Measuring capacitance

Select capacitance measurement (⊣⊦). Connect test leads to the V and COM inputs. Capacitors should be isolated from the circuit to provide accurate DMM measurements (Figure 3-6). Discharge large filter capacitors before attempting to measure them.

Frequency measurements

Select frequency measurement (Hz). Connect test leads to the V and COM inputs. As with other DMM measurements, start at the highest band and switch down to the correct frequency range.

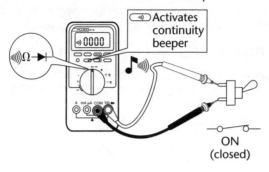

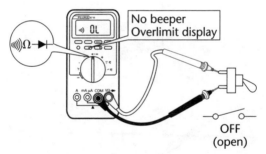

3-5 Testing for continuity.

Testing diodes

Select diode test (→⊢). Connect test leads to the V and COM inputs. Some DMMs have an audible tone for the diode test. Touch the red probe to the anode and the black test probe to the cathode terminal of the diode. The cathode may be marked with a black or white line at one end of the diode (Figure 3-7). A normal silicon diode reading will indicate only an overlimit measurement (OL or 1) if the test leads are reversed.

Digital Multimeter Safety Features

Hand-held test meters should never be connected to any electrical equipment or system operating at a

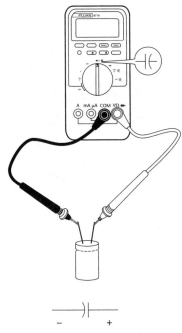

3-6 Measuring capacitance.

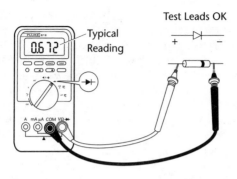

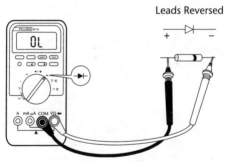

3-7 Testing diodes.

voltage that exceeds the meter's rating. While this is an important safety precaution when using any meter, it becomes even more important with DMMs. Digital meters are more sensitive than older analog models to transient overvoltages caused by nearby lightning strikes, utility switching, motor starting, and capacitor switching. High-voltage transients can damage the electronic circuitry inside DMMs, and in severe cases cause meters to explode.

DMMs have internal fuses which function to protect the test instrument (and the person using it) from harm when taking readings on systems of higher voltage or current rating than the DMM. *However, it is still extremely important never to try to take a reading on a system whose voltage or current is higher than the rating of the DMM itself.*

The International Electrotechnical Commission (IEC) has established four energy-rating categories for test and measurement equipment, with CAT IV offering the highest level of protection.

CAT IV covers utility connections and all outdoor conductors (because of lightning hazards). Examples include service entrance equipment, watt-hour meters, and switchboards/switchgear.

CAT III covers power distribution equipment within buildings and similar structures. This includes panelboards, feeders, busways, motors, and lighting.

CAT II covers single-phase, receptacle-connected loads located more than 10 meters (33 feet) from a CAT III power source or more than 20 m (66 ft) from a CAT IV source.

CAT I covers electronic equipment.

Digital multimeters are certified to these four IEC categories by independent testing laboratories. The certification level is marked directly on the DMMs, and often included in advertising for them. Higher-rated meters can safely be used for lower-level measurement functions. IMPORTANT: The category number of a DMM is more important than its voltage rating when determining the degree of protection that it provides. In other words, a CAT III, 600-V meter offers better protection against high-energy transients than a CAT II, 1000-V meter.

General Safety Precautions for Using Digital Multimeters

- When schematic drawings, building plans, or other documentation is available, check for expected ranges of voltage, current, resistance, and other properties before taking measurements with the DMM. Rotate the function switch to the appropriate range.
- If the appropriate range for a given measurement isn't known, start at the highest scale for voltage, current, etc. Select progressively lower ranges until the measurement falls within the correct range. If the overlimit display (OL or 1) comes on, turn to a higher measurement scale.
- Remove test leads from the circuit or device being tested when changing the measurement range.
- Resistance and diode measurements should only be taken in de-energized circuits. Discharge all capacitors before taking capacitance readings with a DMM.

CHAPTER 4

Troubleshooting Basics

Much of the work performed by electricians and technicians involves the repair and maintenance of electrical equipment and systems. To maintain such systems at peak performance, workers must have a good knowledge of what is commonly referred to as troubleshooting—the ability to determine the cause of a malfunction and then correct it.

Troubleshooting covers a wide range of problems from small jobs such as finding a short circuit or ground fault in a home appliance to tracing out defects in a complex industrial installation. The basic principles used are the same in either case. Troubleshooting requires a thorough knowledge of electrical theory and testing equipment, combined with a systematic and methodical approach to finding and diagnosing the problem.

The following general tips and principles are intended to help define the troubleshooting process. Specific types of electrical equipment and systems are described in later chapters of this book.

Think Before Acting

Study the problem thoroughly, and then ask yourself these questions:

- What were the warning signs preceding the trouble?
- What previous repair and maintenance work has been done?
- Has similar trouble occurred before?
- If the circuit, component, or piece of equipment still operates, is it safe to continue operation before further testing?

The answers to these questions can usually be obtained by:

- Questioning the owner or operator of the equipment.
- Taking time to think the problem through.
- Looking for additional symptoms.
- Consulting troubleshooting charts.
- Checking the simplest things first.
- Referring to repair and maintenance records.
- Checking with calibrated instruments.
- Double-checking all conclusions before beginning any repair on the equipment or circuit components

> **Note**
>
> Always check the easiest and obvious things first; following this simple rule will save time and trouble.

The source of many problems can be traced not to one part alone, but to the relationship of one part to another. For instance, a tripped circuit breaker may be reset to restart a piece of equipment, but what caused the breaker to trip in the first place? It could have been caused by a vibrating "hot" conductor momentarily coming into contact with a ground, or a loose connection could eventually cause overheating, or any number of other causes.

Too often, electrically operated equipment is completely disassembled in search of the cause of a certain complaint, and all evidence is destroyed during disassembly operations. Check again to be certain an easy solution to the problem has not been overlooked.

Find and Correct the Cause of Trouble

After an electrical failure has been corrected in any type of electrical circuit or piece of equipment, be sure to locate and correct the cause so the same failure will not be repeated. Further investigation may reveal other faulty components. Also be aware that although troubleshooting charts and procedures greatly help in diagnosing malfunctions, they can never be complete; there are too many variations and solutions for a given problem.

To solve electrical problems consistently, you must first understand the basic parts of electrical circuits, how they function, and for what purpose. If you know that a particular part is not performing its job, then the cause of the malfunction must be within this part or series of parts.

Intermittent Faults

Finding and diagnosing intermittent faults, where a short, open, or other problem occurs only temporarily, or only under certain conditions, is always one of the most difficult troubleshooting operations. Two features found on most digital multimeters can help with identifying intermittent faults.

Continuity capture mode

This feature is useful for finding intermittent connections with small gauge wires and wiring bundles, and even intermittent relay contact. To check for intermittent opens, place the leads across the normally closed or shorted connection and select Continuity Capture mode on the DMM. Wiggle the wire(s) and heat the connection with a heat gun, or cool it with circuit cooler to make the intermittent open appear. When the open in captured (as short as 250 microseconds), the display shows a transition from open to a short.

Intermittent shorts can be found the same way, by connecting to a normally open circuit and using the wiggling and heating/cooling techniques to capture the short. The only difference is that the transition lines will go from the bottom of the display to the top.

Recording mode

Sometimes intermittent faults cannot be successfully induced while observing the DMM display. Some higher-end units have a recording mode with a date and time stamp. This type of DMM can be left connected to a circuit or piece of electrical equipment for an extended period of time to record the occurrence of an intermittent fault. The date and time of occurrence may provide clues that allow the electrician or technician to trace the cause of the fault (Figure 4-1).

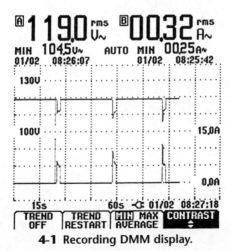

4-1 Recording DMM display.

CHAPTER 5

Troubleshooting Dry-Type Transformers

Dry-type transformers are an essential part of most electrical installations. They range in size from small doorbell transformers to three-phase 25-kVA units suitable for installation in electrical closets (Figure 5-1) to large, free-standing units rated at several hundred kVA (Figure 5-2). Electricians must know how to test for and diagnose problems that develop in transformers—especially in the smaller, dry-type power-supply or control transformers.

Open Circuit: Should one of the windings in a transformer develop a break or "open" condition, no current can flow and therefore, the transformer will not deliver any output. The symptom of an open-circuited transformer is that the circuits which derive power from the transformer are de-energized or "dead." Use an ac voltmeter or DMM to check across the transformer output terminals as shown in Figure 5-3. A reading of 0 V indicates an open circuit.

5-1 Dry-type transformer (25-kVA, three-phase).
(Courtesy of Square D Company.)

5-2 Dry-type transformer (300-kVA, three-phase).
(Courtesy of Square D Company.)

5-3 Checking for an open circuit in a transformer.

Then take a voltage reading across the input terminals. If a voltage reading is present, then the conclusion is that one of the windings in the transformer is open.

However, if no voltage reading is on the input terminals either, then the conclusion is that the open is elsewhere on the line side of the circuit; perhaps a disconnect switch is open.

WARNING!

Make absolutely certain that your testing instruments are designed for the job and are calibrated for the correct voltage. Never test the primary side of any transformer over 600 V unless you are qualified, have the correct high-voltage testing instruments, and the test is made under the proper supervision.

However, if voltage is present on the line or primary side and no voltage is on the secondary or load side, open the switch to de-energize the circuit, and place a warning tag (tag-out and lock) on this switch so that it is not inadvertently closed again while someone is working on the circuit. Disconnect all of the transformer primary and secondary leads and check each winding in the transformer for continuity (a continuous circuit), as indicated by a resistance reading taken with an ohmmeter.

Continuity is indicated by a relatively low resistance reading on control transformers, while an open winding will be indicated by an infinite resistance reading (OL or 1). In most cases, such small transformers will have to be replaced, unless of course the break is accessible and can be repaired.

Ground Fault: Sometimes a few turns in the secondary winding of a transformer will acquire a partial short, which in turn will cause a voltage drop across the secondary. The symptom of this condition is usually overheating of the transformer caused by large circulating currents flowing in the shorted windings.

The easiest way to check this condition is with a voltmeter. Take a reading on the line or primary side of the transformer first to make certain normal voltage is present. Then take a reading on the secondary side. If the transformer has a partial short or ground fault, the voltage reading should be lower than normal.

Replace the faulty transformer with a new one and again take a reading on the secondary. If the voltage reading is now normal and the circuit operates satisfactorily, leave the replacement transformer in the circuit, and either discard or repair the original transformer.

A DMM or ohmmeter may also be used to test this condition when the system is de-energized and the leads are disconnected; a lower resistance reading than normal indicates this condition. However, the reading will usually be so slight that the average ohmmeter is not sensitive enough to detect the difference. Therefore, the recommended way is to use the voltmeter test.

Complete Short: Occasionally a transformer winding will become completely shorted. In most cases, this will activate the overload-protective device and de-energize the circuit, but in other instances, the transformer may continue trying to operate with excessive overheating—due to the very large circulating current. This heat will often melt the wax or insulation inside the transformer, which is easily detected by the odor. Also, there will be no voltage output across the shorted winding and the circuit across the winding will be dead.

The short may be in the external secondary circuit or it may be in the transformer's winding. To determine its location, disconnect the external secondary circuit from the winding and take a reading with a voltmeter. If the voltage is normal with the external circuit disconnected, then the problem lies within the external circuit. However, if the voltage reading is still zero across the secondary leads, the transformer is shorted and will have to be replaced.

Grounded Windings: Insulation breakdown is quite common in older transformers—especially those that have been overloaded. At some point, the insulation breaks or deteriorates and the wire becomes exposed. The exposed wire often comes into contact with the transformer housing and grounds the winding.

If a winding develops a ground, and a point in the external circuit connected to this winding is also grounded, part of the winding will be shorted out. The symptoms will be overheating, which is usually detected by feel or smell, and a low voltage reading as indicated on a voltmeter scale. In most cases, transformers with this condition will have to be replaced.

A megohmmeter is the best test instrument to check for this condition. Disconnect the leads from both the primary and secondary windings. Tests can then be performed on either winding by connecting the megger negative test lead to an associated ground and the positive test lead to the winding to be measured.

Insulation resistance should then be measured between the windings themselves. This is accomplished by connecting one test lead to the primary and the second test lead to the secondary. All such tests should be recorded on a record card under proper identifying labels.

The troubleshooting chart in Figure 5-4 covers the most common dry-type transformer problems.

Malfunction	Probable Cause
Overheating	Continuous overload; wrong external connections; poor ventilation; high surrounding air temperatures
	High input voltage
	Clogged air ducts or inadequate ventilation
Reduced to zero voltage	Short turns; loose connections to transformer terminal board
Excess secondary voltage	Input voltage high; dirt accumulations on primary terminal board
High conductor loss	Overload; terminal boards not on identical tap position
Coil distortion	Coils short-circuited
Insulation failure	Continuous overloads; dirt accumulations on coils; mechanical damage in handling; lightning surge
	Very high core temperature due to high input voltage or low frequency

5-4 Troubleshooting chart for dry-type transformers.

Malfunction	Probable Cause
Breakers or fuses open	Short circuit; overload
Excessive cable heating	Improperly bolted connection
High voltage to ground	Usually a static charge condition
Vibration and noise	Low frequency; high-input voltage; core clamps loosened in shipment or handling; loose hardware on enclosure; location
High exciting current	Low frequency; high input voltage; shorted turns (windings)
High core loss	Low frequency; high input voltage
Smoke	Insulation failure
Burned insulation	Lightning surge; switching or line disturbance; broken bushings

5-4 Troubleshooting chart for dry-type transformers.
(Continued)

CHAPTER 6

Troubleshooting Fluorescent Fixtures

The fluorescent lamp (Figure 6-1) is an electrical discharge lighting source in which a mercury arc generates ultraviolet energy, which in turn activates the phosphor coatings on the inside of the tube to produce light. The lamp takes a variety of shapes, the dominant one being a smooth, long, tubular shape of various diameters and lengths. Due to their negative resistance, fluorescent lamps require a ballast to start and limit the current flow through them. In all cases, the lamps in a fluorescent fixture must match the ballast, and vice versa.

Preheat Lamps

Preheat lamps have bipin bases as shown in Figure 6-1 and are manufactured in sizes from 4 to 100 W and in tube lengths from 6 to 60 in long. Either manual or automatic starters are required to produce a separate cycle for cathode heating prior to operation. Some preheat lamps may be operated on *trigger start* circuits with *rapid start* starting characteristics. In either case, lamp operating characteristics are based upon their operation with ballasts.

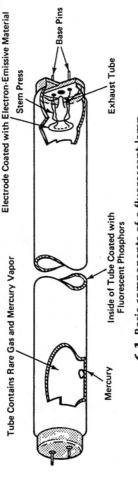

6-1 Basic components of a fluorescent lamp.

Rapid Start Lamps

Rapid start lamps require a preheat cycle, but unlike preheat lamps, do not require a separate starter. The types of lamps falling under this classification are as follows:

- *Rapid start bipin*—limited to 48-in, 34-W T-8 configuration; operates at current of 430 mA.
- *High-output*—recessed double contact available in the range of 24 to 96 in and from 35 to 110 W; operates at current of 800 mA.
- *Power groove*—recessed double contact lamp bases available in this range of 48 to 96 in, and from 110 to 215 W in T-17 bulb size; operates at current of 1500 mA.
- *VHO (very high output) and SHO (super high output)*—manufacturers' brand names for recessed double contact lamp bases in the range of 48 to 96 in, from 110 to 215 W in T-12 bulb size; operates at current of 1500 mA.

Every fluorescent lamp and starter should be tested before installation because lamps and starters are fragile and may possibly be damaged in shipment or handling. This will reveal any defectives and will prevent wasted time in installation, or in maintaining equipment in service. The troubleshooting chart (Figure 6-2) lists faults, probable causes, and corrective action to take while troubleshooting fluorescent lighting fixtures and lamps.

Symptoms	Probable Cause	Action or Items to Check
Blinking on and off, along with shimmering effect during lighted period.	Normal end of lamp life, emission material on electrode depleted.	Replace lamp.
Blinking of relatively new lamps	Incorrect or defective starter. Loose circuit contact.	Replace starter. Seat lamp securely; indicator bumps should be directly over socket slot. Check if lamp holders are rigidly mounted and properly spaced; tighten all connections.

6-2 Troubleshooting chart for fluorescent lamp equipment.

Symptoms	Probable Cause	Action or Items to Check
Blinking with two-lamp ballasts. One lamp starts and one end of the other may blink on and off without starting; eventually both lamps may start.	Cold drafts hitting lamp. Starter leads improperly wired.	Enclose or protect lamp. Check wiring diagram on ballast and reconnect leads correctly or interchange lamp holders at one end of fixture.
Ends of lamp remain lighted.	Starter contacts are stuck together.	Replace starter.

6-2 Troubleshooting chart for fluorescent lamp equipment. *(Continued)*

Symptoms	Probable Cause	Action or Items to Check
No starting effort, or slow starting.	Open circuit in electrodes or air leak in lamp.	Check lamps in another fixture; if no fluorescent end glow exists, replace them.
Slow starting of preheat type lamps.	Starter sluggish or at end of life. Possible open circuit. No starting compensator in leading circuit of two-lamp ballast. Low ballast rating. Low circuit voltage.	Replace starter. Check wiring. Install one in series with starter. Replace with ballast of proper rating. Check voltage and correct if possible.

6-2 Troubleshooting chart for fluorescent lamp equipment. (*Continued*)

Symptoms	Probable Cause	Action or Items to Check
Slow starting of preheat type lamps.	Remote possibility of open-circuited ballasts.	Check ballast.
Slow starting of rapid start type lamps during conditions of high humidity.	Dust or dirt on lamps overcomes effect of silicone coating.	Wash lamps in water containing mild detergent. Rinse lamps with clean water to prevent film from depositing on bulb.
Very slow life accompanied by severe end blackening on rapid start type lamps.	Electrode not being heated.	Check for poor contact between lamp pins and socket. Check for open circuit — loose socket contacts or break in line.

6-2 Troubleshooting chart for fluorescent lamp equipment. *(Continued)*

Symptoms	Probable Cause	Action or Items to Check
Very short lamp life accompanied by severe end blackening on rapid start type lamps.	Loose circuit contact causing on-off blink.	Be sure the lamp holders are rigidly mounted and that the lamp is securely seated; check the circuit wiring.
	Too low or too high voltage.	Check the line voltage and be sure it is within the range on the ballast nameplate.
	Low ambient temperature.	Use special auxiliaries for temperature below 50°F.
	Ballast prematurely starting preheat lamps.	Replace starter with a thermal type.

6-2 Troubleshooting chart for fluorescent lamp equipment. *(Continued)*

Symptoms	Probable Cause	Action or Items to Check
Very short lamp life accompanied by severe end blackening on rapid start type lamps.	No starter compensator in leading circuit of two-lamp ballast.	Install one in series with starter.
	In series instant-start ballasts, one lamp burned out and the other burning dimly. If the burned out lamp is not replaced, the dim lamp will burn out shortly.	Replace the burned out lamp immediately.
	Series instant-start ballasts, both lamps out, only one lamp may be defective.	Check both lamps to determine which is defective; reinstall good lamp.

6-2 Troubleshooting chart for fluorescent lamp equipment. *(Continued)*

Symptoms	Probable Cause	Action or Items to Check
Very short lamp life accompanied by severe end blackening on lamps.	Ballast improperly designed or not within specifications or wrong ballast being used.	Use CBM certified ballast of correct rating for lamp size.
	Improper ballast equipment of dc sources.	Check ballast equipment.
Dense blackening at one or both ends of tube, extending 2 to 3 in. from base.	Normal end of lamp life.	Replace lamp.

6-2 Troubleshooting chart for fluorescent lamp equipment. *(Continued)*

Symptoms	Probable Cause	Action or Items to Check
Dense blackening at one or both ends of tube, extending 2 to 3 in. from base.	With rapid start lamps, accompanied by short life — poor contact between lamp pins and socket.	Check for proper socket spacing or poor socket construction not providing proper wiping of pins when lamp is installed.
Blackening generally within 1 in. of ends.	Mercury deposit.	Should evaporate as lamp is operated.
Blackening early in life.	Starter defective, causing on-off blink or prolonged flashing.	Replace starter.

6-2 Troubleshooting chart for fluorescent lamp equipment. *(Continued)*

Symptoms	Probable Cause	Action or Items to Check
Blackening early in life.	Ends of lamp remain lighted because of starter failure.	Replace starter.
	Too low or too high voltage.	Check the line voltage to be certain it is within the range shown on the ballast nameplate.
	Loose circuit contact causing on-off blink.	Be sure the lamp holders are rigidly mounted and the lamp is securely seated; check circuit wiring.
	No starting compensator in leading circuit.	Install one in series with starter.

6-2 Troubleshooting chart for fluorescent lamp equipment. *(Continued)*

Symptoms	Probable Cause	Action or Items to Check
Blackening early in life.	Ballast improperly designed or not within specifications or wrong ballast being used.	Use CBM certified ballast of correct rating for lamp size.
Dense spot — black about ½ in. wide, extending about halfway around the tube, centering about 1 in. from the base.	Normal sign of age in service. If early in life, indicates excessive lamp starting or high operating current.	Check for off rating or ballast or unusually high-current voltage. Ballast may be improperly designed.

6-2 Troubleshooting chart for fluorescent lamp equipment. *(Continued)*

Symptoms	Probable Cause	Action or Items to Check
Rings — brownish ring at one end or both, about 2 in. from base.	This may develop on some lamps during operation.	Will not affect the lamp performance.
Dark streaks lengthwise on tube.	Globules of mercury condensed on lower part of tube.	The lower half of the tube is cooler than the upper half; by rotating the tube 180° these mercury globules should evaporate due to the increased warmth.

6-2 Troubleshooting chart for fluorescent lamp equipment. *(Continued)*

Symptoms	Probable Cause	Action or Items to Check
Pronounced swirling, spiraling, or fluttering of arc stream.	May occur in new lamps.	Should season out in normal operation.
	Starter not performing properly to preheat electrodes.	Replace starter.
	No starting compensator in leading circuit of two-lamp ballast.	Install one in series with starter.
	Ballast improperly designed or not within specifications or wrong ballast used.	Use CBM certified ballasts of correct rating for lamp size.

6-2 Troubleshooting chart for fluorescent lamp equipment. *(Continued)*

Symptoms	Probable Cause	Action or Items to Check
Pronounced swirling, spiraling, or fluttering of arc stream.	High-voltage starting.	Check voltage and correct, if possible. If condition persists with good operating condition, replace lamp.
Radio interference.	Lamp-radiation broadcasts through radio receiver.	Small condenser in starter or ballast may be defective; replace.
	Line radiation and line feedback.	Apply radio-interference filter at lamp or fixture.

6-2 Troubleshooting chart for fluorescent lamp equipment. *(Continued)*

Symptoms	Probable Cause	Action or Items to Check
Noise: humming sound that may be steady or come and go.	Internal variation in ballast.	Tighten fixture louvers, glass, side panels, etc. If ballast continues to be noisy, replace it.
	Overheated ballasts.	Prolonged blinking of lamp tends to heat ballast; replace ballast.
Dark section of tube — ¼ to ½ of tube gives no light.	DC operation without using reversing switches.	Install reversing switches.
Decreased light output.	Cold drafts hitting tube.	Enclose or protect lamp. Better ventilation of fixture.

6-2 Troubleshooting chart for fluorescent lamp equipment. *(Continued)*

Symptoms	Probable Cause	Action or Items to Check
Decreased light output.	Low-temperature operation.	Enclose the lamp.
	Low circuit voltage.	Check voltage and correct if possible.
	Dust or dirt on lamp, fixture, walls, or ceilings.	Clean.
Lamps operate at unequal brilliancy.	Low circuit voltage on two-lamp lead-lag ballasts.	Check voltage and correct if possible. Possibly defective ballast.
Different color appearance in difference areas of installation.	Holders spaced wrong.	Correct fixture dimensions.

6-2 Troubleshooting chart for fluorescent lamp equipment. *(Continued)*

CHAPTER 7

Troubleshooting Incandescent Lamps and Fixtures

Incandescent lamps are made in thousands of different types and colors from a fraction of a watt to over 10 kW each, and for practically any conceivable lighting application.

Regardless of the type or size, all incandescent filament lamps consist of a sealed glass envelope containing a filament. Light is produced when the filament is heated to incandescence (white glow) by its resistance to a flow of electric current. Figure 7-1 shows the basic components of an incandescent lamp.

The quartz-iodine tungsten-filament lamp is basically an incandescent lamp, since light is produced from the incandescence of its coiled tungsten filament. However, the quartz lamp envelope is filled with an iodine vapor, which prevents the evaporation of the tungsten filament. This evaporation is what normally occurs in conventional incandescent lamps; then the

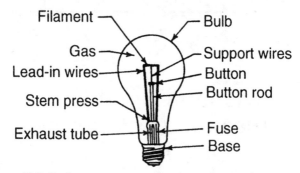

7-1 Basic components of an incandescent lamp.

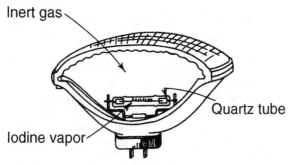

7-2 Basic components of a quartz-iodine tungsten-filament lamp.

bulb begins to blacken, light output deteriorates, and eventually the filament burns out.

While the quartz-iodine lamp has approximately the same efficiency as an equivalent conventional incandescent lamp, it has the advantages of double the normal life, low lumen depreciation, and a smaller bulb for a given wattage. Figure 7-2 illustrates the basic components of a quartz-iodine lamp.

The troubleshooting charts to follow (Figure 7-3) cover the most commonly encountered problems with incandescent lamps and lighting fixtures.

Symptoms	Probable Cause	Action or Items to Check
Lamp not burning, but apparently okay.	Lamp loose.	Tighten in socket.
	Loose or broken connections.	Secure terminals.
		Repair wiring.
Lamp burns dim.	Low voltage.	Match lamp rating to line voltage.
		Increase line voltage.
Short lamp life.	High voltage.	Match lamp rating to line voltage.
	Bulb cracked due to mechanical shock.	Replace lamp.
		Make sure that water does not drip on bulb.

7-3 Troubleshooting chart for incandescent lamp equipment.

Symptoms	Probable Cause	Action or Items to Check
Short lamp life.	Incorrect lamp.	Replace with lamp of size for which luminaire is rated.
	Excessive vibration.	Use shock absorbing device.
Lamp breakage.	Water contacts lamp bulb.	Use enclosed, vaportight luminaire if water spray is present.
		Seal joint where conduit stem enters luminaire.
	Bulb touches luminaire.	Use correct size of lamp.
		Straighten socket.

7-3 Troubleshooting chart for incandescent lamp equipment. *(Continued)*

CHAPTER 8

Troubleshooting HID Lamps and Fixtures

HID (high-intensity discharge) lamp is a generic term which describes a wide variety of lighting sources. All types of HID lamps consist of gaseous discharge arc tubes which, in the versions designed for lighting, operate at pressures and current densities sufficient to generate desired quantities of radiation within their arcs alone.

Mercury vapor lamps contain arc tubes which are formed of fused quartz and radiate ultraviolet energy as well as light. Mercury lamps have outer bulbs that are internally coated with fluorescent materials which, when activated by the ultraviolet, emit visible energy at wavelengths that modify the color of light from the arc.

Multivapor lamps generate light with more than half again the efficiency of the mercury arc, and with better color quality.

The outer bulbs of HID lamps are designed to provide, as nearly as possible, optimum internal environments for arc-tube performance. For example, the rounded shapes labeled E and BT in Figure 8-1 were

8-1 The rounded, shaped bulbs labeled E and BT were devised to maintain uniform temperatures of the bulb walls for better performance of phosphor coatings.

devised to maintain uniform temperatures of the bulb walls for better performance of phosphor coatings. The E bulb improves manufacturing efficiency and eliminates the clear bulb end on phosphor-lined bulbs.

In some cases, special considerations dictate the bulb shape. The R and PAR contours have been selected to achieve desired directional distribution of light. T bulbs are often used in specialized lighting fixtures such as music-stand lamps or fixtures used to illuminate framed paintings and prints.

The troubleshooting chart in Figure 8-2 lists troubleshooting techniques for most mercury and other HID lamp types, along with related luminaires.

Symptoms	Probable Cause	Action or Items to Check
Lamp fails to start or glows feebly.	Lamp has reached end of life.	Replace lamp.
	Lamp was too hot from previous operation.	Relight when lamp has cooled.
	Lamp loose.	Tighten in socket.
		If solder has melted or base eyelet is badly pitted, check for poor contact or defective socket.
	Low temperature.	Make sure that the ballast has adequate voltage to start the lamp at the lowest ambient temperature.

8-2 Troubleshooting chart for HID lamps and related fixtures.

Symptoms	Probable Cause	Action or Items to Check
Lamp fails to start or glows feebly.	Low voltage.	Check open-circuit voltage at socket or ballast tap if incorrectly matched to supply voltage.
		Increase line voltage if necessary.
		Look for wiring fault or poor connections.
Low light output.	Lamp is near end of life.	Replace lamp.
	Low voltage.	Check line voltage and ballast tap selected.
	Wiring fault.	Check wiring and connections.

8-2 Troubleshooting chart for HID lamps and related fixtures. *(Continued)*

Symptoms	Probable Cause	Action or Items to Check
Low light output.	Wrong ballast.	Be sure that ballast is right for the lamp.
	Low output ballast.	Check to see if ballast delivers proper starting current.
	Excessive draft.	Protect single-tube lamps from excessive drafts.
	Dirty, corroded, or inadequate fixture.	Clean, polish, or replace fixture.
	Wrong burning position.	Some fixtures are designed for operation only with their bases up or down. Observe correct burning position.

8-2 Troubleshooting chart for HID lamps and related fixtures. *(Continued)*

Symptoms	Probable Cause	Action or Items to Check
Lamps go out frequently for several minutes at a time.	Voltage dips.	Separate lighting circuit from heavy power circuits.
		Provide voltage regulator.
		Use ballast affording greater protection against voltage dips, such as a regulated output ballast.
	Wiring fault.	Check wiring.
		Tighten connections.
	Frequent power interruptions.	Consult utility company.
	Wrong ballast.	Install correct ballast.

8-2 Troubleshooting chart for HID lamps and related fixtures. *(Continued)*

Symptoms	Probable Cause	Action or Items to Check
Outer bulb color abnormally greenish or yellowish.	Lamp is worn out from exceptionally long service, so that arc tube has discolored.	Replace lamp.
Internal parts oxidized.	Outer bulb seal destroyed.	Replace lamp.
	Dirt or dust.	Wipe bulb carefully so as not to scratch surface. Clean fixture.
Annoying stroboscopic effect.	Cyclic flicker.	Connect fixtures in a staggered arrangement on a three-phase supply.
		Use a two-lamp lead-lag ballast.

8-2 Troubleshooting chart for HID lamps and related fixtures. *(Continued)*

Symptoms	Probable Cause	Action or Items to Check
Annoying stroboscopic effect.	Cyclic flicker.	Add incandescent lamps to the system.
Radio interference.	Circuit components.	Check circuit, as lamp itself generally creates no radio interference, and standard ballast, usually suppress any line re-radiation.
		If individual lamp ballasts are not used, it may be necessary to add a small capacitor across the lamp.
Noisy ballast	Cyclic vibration.	Tighten ballast cover.

8-2 Troubleshooting chart for HID lamps and related fixtures. *(Continued)*

Symptoms	Probable Cause	Action or Items to Check
Noisy ballast.	Cyclic vibration.	Replace ballast.
Sunburn or suntan.	Exposure to rays of lamp without outer bulb or prolonged exposure at very short distance from lamp.	Replace lamps that lack an outer bulb. Turn lights out when working near them or cover exposed parts of body. Use enclosed fixtures.
Lamp won't enter fixture.	Fixture opening too small or socket off center.	Relocate socket or try a socket extender if change in position of light center is not critical.
	Damaged base.	Replace base. Check for rough handling.

8-2 Troubleshooting chart for HID lamps and related fixtures. *(Continued)*

Symptoms	Probable Cause	Action or Items to Check
Lamp won't enter fixture.	Crooked base.	Check base-bulb alignment. If more than 3° in any direction, lamp may be defective.
Overcurrent devices open when lamps are started.	High transient current of very short duration. Usually caused by ballast or circuit components.	Use time-delay fused or thermal-magnetic type circuit breakers.
Lamp breakage or cracks in outer bulb.	Shipping damage or mishandling.	Check carrier and workers handling bulbs.
	Water drips on hot bulb.	Check for leaks or for condensation in fixture.
	Poorly sealed fixtures breathe in moist air.	Use enclosed vapor-tight fixture.

8-2 Troubleshooting chart for HID lamps and related fixtures. *(Continued)*

Symptoms	Probable Cause	Action or Items to Check
Lamp breakage.	Bulb touches fixture, edge of sockets, metallic bulb changer, heat conductor, or heat insulator during insertion or operation.	Straighten or adjust socket.
		Use socket extender to provide ample clearance.
		Remove metallic or other element touching bulb.
		Use weather duty lamps.
	Bulb cracked through mechanical shock.	Replace lamp.
		Do not allow bulbs to bump against each other.
Arc tube swollen, cracked, or broken.	Overwattage operation.	Check ballast and socket wiring.

8-2 Troubleshooting chart for HID lamps and related fixtures. *(Continued)*

Symptoms	Probable Cause	Action or Items to Check
Arc tube swollen, cracked, or broken.	Wrong burning position.	Correct burning position.
	Incorrect magnet polarity or magnet ineffective.	Correct magnet polarity or replace magnet or use lamps without magnets.
Ends of arc tube split apart or internal wires melted.	Excessive current or voltage due to operation without ballast or else lightning damage.	Check ballast and socket wiring.
		Check for possibility of lightning damage.
		Check for very high voltage applied to a hot lamp, even momentarily, resulting in discharge in outer tube.

8-2 Troubleshooting chart for HID lamps and related fixtures. *(Continued)*

Symptoms	Probable Cause	Action or Items to Check
Loose base.	Abnormal operation or defective lamp.	Check for operation beyond published temperature limits.
		Check if basing cement was severely disturbed prior to or within first 10 hours of operation.
Rattle.	Piece of loose glass or cement.	Lamps having this condition should not be considered defective if they are otherwise satisfactory.
Blackening of inner arc tube.	Long service.	Replace lamp if lamp output is too low.
	Overwattage operation.	Adjust.

8-2 Troubleshooting chart for HID lamps and related fixtures. *(Continued)*

CHAPTER 9

Troubleshooting Electric Motors

Electric motors utilize the principle of electromagnetic induction. An induction motor has a stationary part, or stator, with windings connected to the ac supply, and a rotating part, or rotor, which contains coils or bars. There is no electrical connection between the stator and rotor. The magnetic field produced in the stator windings induces a voltage in the rotor coils or bars.

When an induction motor malfunctions, the stator (stationary) windings will usually be defective, and these windings will then have to be repaired or replaced. Stator problems can usually be traced to one or more of the following causes:

- Worn bearings
- Moisture
- Overloading
- Operating single phase
- Poor insulation

Dust and dirt are usually contributing factors. Some forms of dust are highly conductive and contribute materially to insulation breakdown. The effect of dust on the motor temperature through restriction of ventilation is another reason for keeping the machine clean, either by periodically blowing it out with compressed air or by dismantling and cleaning. The compressed air must be dry and throttled down to a low pressure that will not damage the insulation.

One of the worst enemies of motor insulation is moisture. Therefore, motor insulation must be kept reasonably dry, although many applications make this practically impossible unless a totally enclosed motor is utilized. If a motor must be operated in a damp location, a special moisture-resisting treatment should be given to the windings.

The life of a winding depends upon keeping it in its original condition as long as possible. In a new machine, the winding is snug in the slots and the insulation is fresh and flexible. This condition is best maintained by periodic cleaning, followed by varnish and oven treatments.

After insulation dries out, it loses its flexibility and the mechanical stresses caused by starting and plugging, as well as the natural stresses in operation under load, will tend to cause short circuits in the coils and possibly failures from the coil to ground—usually at the point where the coil leaves the slot. The effect of periodic varnish and oven treatments properly carried out so as to fill all air spaces caused by drying and shrinkage of the insulation will maintain a solid winding, and also provide an effective seal against moisture.

Troubleshooting Motors

To detect defects in electric motors, the windings are normally tested for ground faults, opens, shorts, and reverses. The exact method of performing these tests will depend on the type of motor being serviced. However, regardless of the motor type, a knowledge of some important terms is necessary before maintenance personnel can approach their work satisfactorily.

Ground: A winding becomes grounded when it makes an electrical contact with the iron of the motor. The usual causes of grounds include the following: Bolts securing the end plates come into contact with the winding; the wires press against the laminations at the corners of the slots, which is likely to occur if the slot insulation tears or cracks during winding; and the centrifugal switch may be grounded to the end plate.

Open circuits: Loose or dirty connections as well as a broken wire can cause an open circuit in an electric motor.

Shorts: Two or more turns of the coil that contact each other electrically will cause a short circuit. This condition may develop in a new winding if the winding is tight and much pounding is necessary to place the wires in position. In other cases, excessive heat developed from overloads will make the insulation defective and will cause shorts. A short circuit is usually detected by observing smoke from the windings as the motor operates or when the motor draws excessive current at no load.

Tools for Troubleshooting

In addition to small portable testing devices such as voltmeters, ammeters, brush-spring tension testers, and a transistorized stethoscope for checking motor bearings, maintenance equipment should include a 500-volt insulation-resistance tester, a spark-gap oil dielectric tester, and a portable oil filtering unit.

Transistorized stethoscope: This type of testing instrument—equipped with a transistor-amplifier—is used to ascertain the condition of motor bearings. A little practice in interpreting what is heard through it may be required, but in general, it is relatively simple to use. if, when the stethoscope is applied to a motor bearing, a purring sound is heard, the bearing is usually normal. On the other hand, a thumping sound or a rough grinding sound indicates a failing bearing.

Insulation-resistance tests: When performing an insulation-resistance test, first make a careful safety check and ensure that all circuits and equipment are rated at the voltage of the megger. Furthermore, all equipment scheduled for testing must be disconnected from all power sources. All safety switches should be opened and locked out to make certain that motor starters or other control equipment cannot accidentally energize the apparatus.

After the readings have been recorded, they are corrected for temperature using a temperature-correction chart supplied with most meggers. As a rule of thumb, most maintenance departments feel that 600-V winding insulation is acceptable if the corrected resistance value is 1 MΩ or more. High resistance readings which show a continuing downward trend over a period of time indicate failing insulation.

The chart in Figure 9-1 gives a list of practical tools and equipment for effective electrical maintenance of motors and for other electrical apparatus as well.

Grounded Coils

The usual effect of one grounded coil in a motor winding is the repeated blowing of a fuse, or tripping of the circuit breaker, when the line switch is closed, that is, provided the machine frame and the line are both grounded. Two or more grounds will give the same result and will also short out part of the winding in that phase in which the grounds occur. A quick and simple test to determine whether or not a ground exists in the winding can be made with a conventional continuity tester. In testing with such an instrument, first make certain that the line switch is open and locked out, causing the motor leads to be *de-energized*. Place one test lead on the frame of the motor and the other in turn on each of the line wires leading from the motor. If there is a grounded coil at any point in the winding, the lamp of the continuity tester will light, or in the case of a meter, the dial will swing toward *infinity*.

To locate the phase that is grounded, test each phase separately. In a three-phase winding it will be

Tools or Equipment	Application
Multimeters, voltmeters, ohmmeters, clamp-on ammeters, wattmeters, clamp-on Power factor meter	Measure circuit voltage, resistance, current and power. Useful for circuit tracing and troubleshooting.
Potential and current transformers, meter shunts	Increase range of test instruments to permit reading of high-voltage and high-current circuits.
Transistorized stethoscope	Detect faulty rotating machinery bearings and leaky valves.
Tachometer	Check rotating machinery speeds.

9-1 Tools for electric motor maintenance.

necessary to disconnect the star or delta connections, if accessible. After the grounded phase is located the pole-group connections in that phase can be disconnected and each group tested separately. When the leads are placed, one on the frame and the other on the grounded coil group, the lamp will indicate the ground in this group by lighting. The stub connections

Tools or Equipment	Application
Recording meters, instruments	Provide permanent record of voltage, current, power, temperature, etc., on charts for analytic study.
Insulation resistance tester, thermometer, psychrometer	Test and monitor insulation resistance; use thermometer and psychrometer for temperature-humidity correction.
Portable oil dielectric tester; portable oil filter	Test OCB, transformer oil, or other insulating oils. Recondition used oil.
Air gap feeler gauges	Check motor or generator air gap between rotor and stator.
Cleaning solvent	Removes grease or dirt from motor windings or other electrical parts.

9-1 Tools for electric motor maintenance.
(Continued)

Tools or Equipment	Application
Hand stones (rough, medium, fine), grinding rig, canvas strip	Grinding, smoothing, and finishing commutators or slip rings.
Spring tension scale	Checks brush pressure on dc motor commutators or on ac motor skip rings; tests electrical contact pressure on relays, starters, or contactors.
Magnifying glass, binoculars	Use magnifying glass to examine brushes, commutators, or small electrical contacts or parts; binoculars allow close inspection of remote or high-voltage parts.
Motor rotation tester	Checks direction of motor rotation before connection.

9-1 Tools for electric motor maintenance. *(Continued)*

between the coils and this group may then be disconnected and each coil tested separately until the exact coil that is grounded is located.

Sometimes moisture in the insulation around the coils on old and defective insulation will cause a high-resistance ground that is difficult to detect with a test lamp. A megger can be used to detect such faults.

Armature windings and the commutator of a motor may be tested for grounds in a similar manner. On some motors, the brush holders are grounded to the end plate. Consequently, before the armature is tested for grounds, the brushes must be lifted away from the commutator.

When a grounded coil is located, it should be either removed and reinsulated or cut out of the circuit. At times, however, it may be inconvenient to stop a motor long enough for a complete rewinding or permanent repairs. In such cases, when trouble develops, it is often necessary to make a temporary repair until a later time when the motor may be taken out of service long enough for rewinding or permanent repairs.

To temporarily repair a defective coil, a jumper wire of the same size as that used in the coils is connected to the bottom lead of the coil immediately adjacent to the defective coil and run across to the top lead of the coil on the other side of the defective coil, leaving the defective coil entirely out of the circuit. The defective coil should then be cut at the back of the winding and the leads taped so as not to function when the motor is started again. If the defective coil is grounded, it should also be disconnected from the other coils.

Shorted Coils

Shorted turns within coils are usually the result of failure of the insulation on the wires. This is frequently caused by the wires being crossed and having excessive pressure applied on the crossed conductors when the coils are being inserted in the slot. Quite often it is caused by using too much force in driving the coils down in the slots. In the case of windings that have been in service for several years, failure of the insulation may be caused by oil, moisture, etc. If a shorted coil is left in a winding, it will usually burn out in a short time and if it is not located and repaired promptly will probably cause a ground and the burning out of a number of other coils.

One inexpensive way of locating a shorted coil is by the use of a growler and a thin piece of steel. Figure 9-2 shows a sketch of a growler in use in a stator. Note that the poles are shaped to fit the curvature of the teeth inside the stator core. The growler should be placed in the core as shown, and the thin piece of steel should be placed the distance of one coil span away from the center of the growler. Then, by moving the growler around the bore of the stator and always keeping the steel strip the same distance away from it, all of the coils can be tested.

If any of the coils has one or more shorted turns, the piece of steel will vibrate very rapidly and cause a loud humming noise. By locating the two slots over which the steel vibrates, both sides of the shorted coil can be found. If more than two slots cause the steel to vibrate, they should all be marked, and all shorted

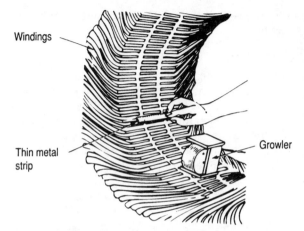

9-2 Growler used to test a stator of an ac motor.

coils should be removed and replaced with new ones or cut out of the circuit as previously described.

Sometimes one coil or a complete coil group becomes short-circuited at the end connections. The test for this fault is the same as that for a shorted coil. If all the coils in one group are shorted, it will generally be indicated by the vibration of the steel strip over several consecutive slots, corresponding to the number of coils in the group.

The end connections should be carefully examined, and those that appear to have poor insulation should be moved during the time that the test is being made. It will often be found that when the shorted end connections are moved during the test the vibration of the steel will stop. If these ends are reinsulated, the trouble should be eliminated.

Open Coils

When one or more coils become open-circuited by a break in the turns or a poor connection at the end, they can be tested with a continuity tester as previously explained. If this test is made at the ends of each winding, an open can be detected by the lamp failing to light. The insulation should be removed from the pole-group connections, and each group should be tested separately.

An open circuit in the starting winding may be difficult to locate, since the problem may be in the centrifugal switch as well as the winding itself. In fact, the centrifugal switch is probably more apt to cause trouble than the winding since parts become worn, defective, and more likely, dirty. Insufficient pressure of the rotating part of centrifugal switches against the stationary part will prevent the contacts from closing and thereby produce an open circuit.

If the trouble is a loose connection at the coil ends, it can be repaired by resoldering the splices, but if it is within the coil, the coil should either be replaced or a jumper should be connected around it until a better repair can be made.

Reversed Connections

Reversed coils cause the current to flow through them in the wrong direction. This fault usually manifests itself — as do most irregularities in winding connections—by a disturbance of the magnetic circuit, which results in excessive noise and vibration. The fault can be located by the use of a magnetic compass and some source of low-voltage direct current. This voltage should be adjusted so it will send about one-fourth to one-sixth of the full-load current through the winding, and the dc leads should be placed on the start and finish of one phase. If the winding is three-phase, star-connected, this would be at the start of one phase and the star point. If the winding is delta-connected, the delta must be disconnected and each phase tested separately.

Place a compass on the inside of the stator and test each of the coil groups in that phase. If the phase is connected correctly, the needle of the compass will reverse definitely as it is moved from one coil group to another. However, if any one of the coils is reversed, the reversed coil will build up a field in the direction opposite to the others, thus causing a neutralizing effect which will be indicated by the compass needle refusing to point definitely to that group. If there are only two coils per group, there will be no indication if one of them is reversed, as that group will be completely neutralized.

When an entire coil group is reversed, it causes the current to flow in the wrong direction in the whole group. The test for this fault is the same as that for reversed coils. The winding should be magnetized with direct current, and when the compass needle is passed around the coil groups, they should indicate alternately N.S., N.S., etc. If one of the groups is reversed, three consecutive groups will be of the same polarity. The remedy for either reversed coil groups or reversed coils is to make a visual check of the connections at that part of the winding, locate the wrong connection, and reconnect it properly.

When the wrong number of coils are connected in two or more groups, the trouble can be located by counting the number of ends on each group. If any mistakes are found, they should be remedied by reconnecting properly.

Reversed Phase

Sometimes in a three-phase winding a complete phase is reversed by either having taken the starts from the wrong coils or by connecting one of the windings in the wrong relation to the others when making the star or delta connections. If the winding is delta-connected, disconnect any one of the points where the phases are connected together and pass current through the three windings in series. Place a compass on the inside of the stator and test each coil group by slowly moving the compass one complete revolution around the stator.

The reversals of the needle in moving the compass one revolution around the stator should be three times the number of poles in the winding.

In testing a star- or wye-connected winding, connect the three starts together and place them on one dc lead. Then connect the other dc lead and star point, thus passing the current through all three windings in parallel. Test with a compass as explained for the delta winding. The result should then be the same, or the reversals of the needle in making one revolution around the stator should again be three times the number of poles in the winding.

These tests for reversed phases apply to full-pitch windings only. If the winding is fractional-pitch, a careful visual check should be made to determine whether there is a reversed phase or mistake in connecting the star or delta connections.

The troubleshooting chart in Figure 9-3 may be used by qualified personnel who have the proper tools and equipment. These instructions do not cover all details or variations in equipment, nor do they provide for every possible condition to be met in actual practice.

Troubleshooting Split-Phase Motors

If a split-phase motor fails to start, the trouble may be due to one or more of the following faults:

- Tight or "frozen" bearings
- Worn bearings, allowing the rotor to drag on the stator
- Bent rotor shaft
- One or both bearings out of alignment
- Open circuit in either starting or running windings
- Defective centrifugal switch
- Improper connections in either winding
- Grounds in either winding or both
- Shorts between the two windings

Symptoms	Probable Cause	Action or Items to Check
Slow speed	Open primary circuit.	Locate fault with testing device and repair.
Slow to accelerate	Excess loading.	Reduce load.
	Poor circuit.	Check for high resistance.
	Defective squirrel-cage rotor.	Replace.
	Applied voltage too low.	Get power company to increase voltage tap.
Wrong rotation	Wrong sequence of phases.	Reverse connections at motor or at switchboard.

9-3 Troubleshooting chart for motors.

Symptoms	Probable Cause	Action or Items to Check
Motor overheats	Check for overload.	Reduce load.
	Wrong blowers or air shields.	May be clogged with dirt and prevent proper ventilation of motor.
	Motor may have one phase open.	Check to make sure that all leads are well connected.
	Grounded coil.	Locate and repair.
	Unbalanced terminal voltage.	Check to make sure that all leads are well connected.
	Grounded coil.	Locate and repair.

9-3 Troubleshooting chart for motors. *(Continued)*

Symptoms	Probable Cause	Action or Items to Check
Motor overheats	Unbalanced terminal voltage.	Check for faulty leads.
	Shorted stator coil.	Repair and then check wattmeter reading.
	Faulty connection.	Indicated by high resistance.
	High voltage.	Check terminals of motor with voltmeter.
	Low voltage.	Same as above.
Motor stalls	Wrong application.	Change type or size. Consult manufacturer.

9-3 Troubleshooting chart for motors. *(Continued)*

Symptoms	Probable Cause	Action or Items to Check
Motor stalls	Overloaded motor.	Reduce load.
	Low motor voltage.	See that nameplate voltage is maintained.
	Open circuit.	Fuses blown.
	Incorrect control resistance of wound rotor.	Check control sequence. Replace broken resistors. Repair open circuits.
Motor does not start	One phase open.	See that no phase is open. Reduce load.
	Defective rotor.	Look for broken bars or rings.

9-3 Troubleshooting chart for motors. *(Continued)*

Symptoms	Probable Cause	Action or Items to Check
Motor does not start	Poor stator coil connection.	Remove end bells.
Motor runs, then quits	Power failure.	Check for loose connections to line, to fuses, and to control.
Slow speed	Not applied properly.	Consult supplier for proper type.
	Voltage too low at motor terminals because of line drop.	Use higher voltage on transformer terminals or reduce load.

9-3 Troubleshooting chart for motors. *(Continued)*

Symptoms	Probable Cause	Action or Items to Check
Slow speed	If wound rotor, improper control operation of secondary.	Correct secondary control.
	Low pull-in torque of synchronous motor.	Change rotor starting resistance or change rotor design.
	Check that all brushes are riding on rings.	Check secondary connections; leave no leads poorly connected.
	Broken rotor bars.	Look for cracks near the rings. A new rotor may be required.

9-3 Troubleshooting chart for motors. *(Continued)*

Symptoms	Probable Cause	Action or Items to Check
Motor vibrates	Motor misaligned.	Realign.
	Weak foundations.	Strengthen base.
	Coupling out of balance.	Balance coupling.
	Driven equipment unbalanced.	Rebalance driven equipment.
	Defective ball bearing.	Replace bearing.
	Bearing not in line.	Line up properly.
	Balancing weights shifted.	Rebalance motor.

9-3 Troubleshooting chart for motors. *(Continued)*

Symptoms	Probable Cause	Action or Items to Check
Motor vibrates	Wound rotor coils replaced.	Rebalance motor.
	Polyphase motor running single-phase.	Check for open circuit.
	Excessive end play.	Adjust bearing or add washer.
Unbalanced line current	Unequal terminal volts.	Check leads and connections.
	Single-phase operation.	Check for open circuit.
	Poor rotor contacts in control wound rotor resistance.	Check control devices.

9-3 Troubleshooting chart for motors. *(Continued)*

Symptoms	Probable Cause	Action or Items to Check
Unbalanced line current	Brushes not in proper position in wound rotor.	See that brushes are properly seated and shunts in good condition.
	Fan rubbing air shield.	Remove interference.
	Fan striking insulation.	Clear fan.
	Loose bedplate.	Tighten holding bolts.
Magnetic noise	Air gap not uniform.	Check and correct bracket fits or bearing.

9-3 Troubleshooting chart for motors. *(Continued)*

Tight or worn bearings: Tight or worn bearings may be due to the lubricating system failing, or when new bearings are installed, they may run hot if the shaft is not kept well oiled.

If the bearings are worn to such an extent that they allow the rotor to drag on the stator, this will usually prevent the rotor from starting. The inside of the stator laminations will be worn bright where they are rubbed by the rotor. When this condition exists, it can generally be easily detected by close observation of the stator field and rotor surface when the rotor is removed.

Bent shaft and bearings out of line: A bent rotor shaft will usually cause the rotor to bind when in a certain position and then run freely until it comes back to the same position again. An accurate test for a bent shaft can be made by placing the rotor between centers on a lathe and turning the rotor slowly while a tool or marker is held in the tool post close to the surface of the rotor. If the rotor wobbles, it is an indication of a bent shaft.

Bearings out of alignment are usually caused by uneven tightening of the end-shield plates. When placing end shields or brackets on a motor, the bolts should be tightened alternately, first drawing up two bolts which are diametrically opposite. These two should be drawn up only a few turns and the other kept tightened an equal amount all the way around. When the end shields are drawn up as far as possible with the bolts, they should be tapped tightly against the frame with a mallet and the bolts tightened again.

Open circuits and defective centrifugal switches: Open circuits in either the starting or running winding will cause the motor to fail to start. This fault can be detected by testing in series with the start and finish of each winding with a test lamp or ohmmeter.

A defective centrifugal switch will often cause considerable trouble that is difficult to locate unless one has good knowledge of the operating characteristics of these switches. If the switch fails to close when the rotor stops, the motor will not start when the line switch is closed. Failure of the switch to close is generally caused by dirt, grit, or some other foreign matter getting into the switch. The switch should be thoroughly cleaned with a degreasing solution such as AWA1,1,1 and then inspected for weak or broken springs.

If the winding is on the rotor, the brushes sometimes stick in the holders and fail to make good contact with the slip rings. This causes sparking at the brushes. There will probably also be a certain place where the rotor will not start until it is moved far enough for the brush to make contact on the ring. The brush holders should be cleaned and the brushes carefully fitted so they move more freely with a minimum of friction between the brush and the holders. If a centrifugal switch fails to open when the motor is started, the motor will probably growl and continue to run slowly, causing the starting winding to burn out if not promptly disconnected from the line. In most cases, however, the "heaters" in the motor control will take care of this before any serious damage occurs. This fault is likely to be caused by dirt or hardened grease in the switch.

Reversed connections and grounds: Reversed connections are caused by improperly connecting a coil or group of coils. The wrong connections can be found and corrected by making a careful check on the connections and reconnecting those that are found at fault. The test with a dc power source and a compass can also be used for locating reversed coils. Test the starting and running windings separately, exciting only one winding at a time, with direct current. The compass should show alternate poles around the winding.

The operation of a motor that has a ground in the winding will depend on where the ground is and whether or not the frame is grounded. If the frame is grounded, then when the ground occurs in the winding, it will usually blow a fuse or trip the overcurrent protective device.

A test for grounds can be made with a test lamp or continuity tester. One test lead should be placed on the frame and the other on a lead to the winding. If there is no ground, the lamp will not light, nor will any deflection be present when a meter is used. If the light does light or the meter shows continuity, it indicates a ground is present—due to a defect somewhere in the motor's insulation.

Short circuits: Short circuits between any two windings can be detected by the use of a test lamp or continuity tester. Place one of the test leads on one wire of the starting winding and the other test lead on the wire of the running winding. If these windings are properly insulated from each other, the lamp should not light. If it does, it is a certain indication that a short or ground fault exists between the windings. Such a condition will usually cause part of the starting winding to burn out. The starting winding is always wound on top of the running winding, so if it becomes defective due to a defective centrifugal switch or a short circuit, the starting winding can be conveniently removed and replaced without disturbing the running winding.

Storing Motors

There are many reasons for storing motors, but the two major ones are:

- The project on which they are to be used is not complete.
- Spare motors are often kept as back-ups on most industrial installations.

The first consideration when storing motors for any length of time is the location. A dry location should be selected if at all possible—one that does not undergo severe changes in temperature over a 24-hour period. When the ambient temperature changes frequently during a 24-hour period, condensation is certain to form on the motor, and moisture is one of the worst enemies of motor insulation. Therefore, guarding against moisture is one of the chief concerns when storing motors of any type.

A means for transporting the motors from the place of storage to the place where they will be used, or else shifted around in the storage area, is also of importance. A motors should not be lifted by its rotating shaft. Doing so can damage the alignment of the rotor in relationship to the stator. Even picking up the smaller fractional horsepower motors by the shaft is not recommended. Many workers have received bad cuts from the sharp keyways on motor shafts when picked up with the bare hands.

When an electric motor is received at the job site, always refer to the manufacturer's instructions and follow them to the letter. Failure to do so could result in serious injury to the workers and motor alike.

Once the motor has been uncrated, check to see if any damage has occurred during handling. Be sure that the motor shaft and armature turn freely. This is also a good time to check to determine if the motor has been exposed to dirt, grease, grit, or excessive moisture in either shipment or storage. Motors in storage should have shafts turned over once each month to redistribute grease in the bearings

Note

Motors in storage should have shafts turned over once each month to redistribute grease in the bearings.

Warning!

Never start a motor which has been wet without first having it thoroughly dried.

The measure of insulation resistance is a good dampness test. Clean the motor of any dirt or grit before putting it back in service.

Eyebolts of lifting lugs on motors are intended only for lifting the motor and factory motor-mounted standard accessories. These lifting provisions should never be used when lifting or handling the motor when the motor is attached to other equipment as a single unit.

The eyebolt lifting-capacity rating is based on a lifting alignment coincident with the eyebolt centerline. The eyebolt capacity reduces as deviation from this alignment increases.

The following is a list of items that must be considered when storing motors for any length of time:

- Make sure motors are kept clean.
- Make sure motors are kept dry.
- Supply supplemental heating in the storage area if necessary.
- Motors should be stored in an orderly fashion, that is, grouped by horsepower, etc.
- Motor armatures should be rotated periodically.
- Lubrication should be checked periodically.
- Protect shafts and keyways during storage and also while transporting motors from one location to another.
- Test motor-winding resistance upon receiving; test again after setting in storage.

Identifying Motors

Electrical workers will sometimes come across a motor with no identification (no nameplate or lead tags) which must be put back into service or else repaired. The experienced electrician should know how to positively identify the motor's characteristics, even with no written data.

The NEMA Standard method of motor identification is easy to remember by drawing the coils to form a wye. Identify one outside coil end with the number one (1), and then draw a decreasing spiral and number each coil end in sequence as shown in Figure 9-4.

By using an ohmmeter, continuity tester, or DMM, the individual circuits can be located as follows.

Step 1. Connect one probe of the tester to any lead, and check for continuity to each of the other eight leads. A reading from only one other lead indicates one of the two wire circuits. A reading to two other leads indicates the three-wire circuit that makes up the internal wye connection.

Step 2. Continue checking and isolating leads until all four circuits have been located. Tag the wires of the three lead circuits T-7, T-8, and T-9 in any order. The other leads should be temporarily marked T-1 and T-4 for one circuit, T-2 and T-5 for the second circuit, and T-3 and T-6 for the third and final circuit.

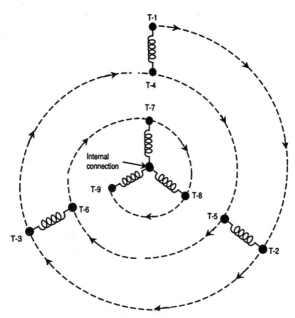

9-4 Identify one outside coil and then draw a decreasing spiral and number each coil.

The following test voltages are for the most common dual-voltage range of 230/460 volts. For other motor ranges, the voltages listed should be changed in proportion to the motor rating.

As all the coils are physically mounted in slots on the same motor frame, the coils will act almost like the primary and secondary coils of a transformer. Figure 9-5 shows a simplified electrical arrangement of the coils. Depending on which coil group power is applied to, the resulting voltage readings will be additive, subtractive, balanced, or unbalanced depending on physical location with regard to the coils themselves.

Step 3. The motor may be started on 230 V by connecting leads T-7, T-8, and T-9 to the three-phase source. If the motor is too large to be connected directly to the line, the voltage should be reduced by using a reduced voltage starter or other suitable means.

Step 4. Start the motor with no load connected and bring up to normal speed.

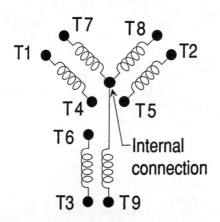

9-5 Simplified electrical arrangement of wye-wound motor coils.

Step 5. With the motor running, a voltage will be induced in each of the open two-wire circuits that were tagged T-1 and T-4, T-2 and T-5, and T-3 and T-6. With a voltmeter, check the voltage reading of each circuit. The voltage should be approximately 125 to 130 V and should be the same on each circuit.

> **Note**
>
> The voltages referred to during the testing are only for reference and will vary greatly from motor to motor, depending on size, design, and manufacturer. If the test calls for equal voltages of 125 to 130 and the reading is only 80 to 90, that is okay as long as the voltage readings are nearly equal.

Step 6. With the motor still running, carefully connect the lead that was temporarily marked T-4 with the T-7 and line lead. Read the voltage between T-1 and T-8 and also between T-1 and T-9. If both readings are of the same value and are approximately 330 to 340 V, leads T-1 and T-4 may be disconnected and permanently marked T-1 and T-4.

Step 7. If the two voltage readings are of the same value and are approximately 125 to 130, disconnect and interchange leads

T-1 and T-4 and mark permanently (original T-1 changed to T-4 and original T-4 changed to T-1).

Step 8. If readings between T-1 and T-8 and between T-1 and T-9 are of unequal values, disconnect T-4 from T-7 and reconnect T-4 to the junction of T-8 and line.

Step 9. Measure the voltage now between T-1 and T-7 and also between T-1 and T-9. If the voltages are equal and approximately 330 to 340 V, tag T-1 is permanently marked T-2 and T-4 is marked T-5 and disconnected. If the readings taken are equal but are approximately 125 to 130 V, leads T-1 and T-4 are disconnected, interchanged, and marked T-2 and T-5 (T-1 changed to T-5, and T-4 changed to T-2). If both voltage readings are different, T-4 lead is disconnected from T-8 and moved to T-9. Voltage readings are taken again (between T-1 and T-7 and T-1 and T-8) and the leads permanently marked T-3 and T-6 when equal readings of approximately 330 to 340 V are obtained.

Step 10. The same procedure is followed for the other two circuits that were temporarily marked T-2 and T-5 and T-3 and T-6, until a position is found where both voltage readings are equal and approximately 330 to 340 V and the tags change to correspond to the standard lead markings as shown in Figure 9-6.

Step 11. Once all leads have been properly and permanently tagged, leads T-4, T-5, and T-6 are connected together and voltage readings are taken between T-1, T-2, and T-3. The voltages should be equal and approximately 230 V.

Step 12. As an additional check, the motor is shut down and leads T-7, T-8, and T-9 are disconnected, and leads T-1, T-2, and T-3 are connected to the line. Connect T-1 to the line lead T-7 was connected to, T-2 to the same line as T-8 was previously connected to, and T-3 to the same lead that T-9 was connected to. With T-4, T-5, and T-6 still connected together to form a wye connection, the motor can again be started without a load. If all lead markings are correct, the motor rotation with leads T-1, T-2, and T-3 connected will be the same as when T-7, T-8, and T-9 were connected.

The motor is now ready for service and is connected in series for high voltage or parallel for low as indicated by the NEMA Standard connections shown in Figure 9-6.

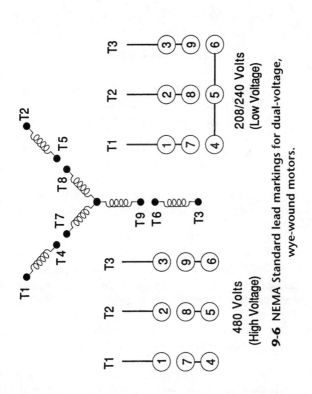

9-6 NEMA Standard lead markings for dual-voltage, wye-wound motors.

> **Note**
> This procedure may not work on some wye-wound motors with concentric coils.

Three-Phase Delta-Wound Motors

Most dual-voltage, delta-wound motors also have nine leads, as indicated in Figure 9-7, but there are only three circuits of three leads each.

Continuity tests are used to find the three coil groups as was done for the wye-wound motor. Once the coil groups are located and isolated, further resistance checks must be made to locate the common wire in each coil group. As the resistance of some delta-wound motors is *very* low, a digital ohmmeter, Wheatstone bridge, or other sensitive device may be needed.

Each coil group consists of two coils tied together with three leads brought out to the motor junction or terminal box. Reading the resistance carefully between each of the three leads shows that the readings from one of the leads to each of the other two leads will be the same (equal), but the resistance reading between those two leads will be double the previous readings; Figure 9-8 may help clarify the technique.

The common lead found in the first coil group is permanently marked T-1, and the other two leads temporarily marked T-4 and T-9. The common lead of the next coil group is found and permanently marked T-2 and the other leads temporarily marked T-5 and

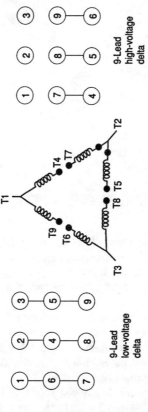

9-7 NEMA Standard lead markings for dual-voltage, delta-wound motors.

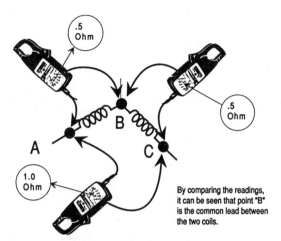

9-8 Using ohmmeter to test motor leads.

T-7. The common lead of the last coil group is located and marked T-3 with the other leads being temporarily marked T-6 and T-8.

After the leads have been marked, the motor may be connected to a 230-V three-phase line using leads T-1, T-4, and T-9. Lead T-7 is connected to line and T-4, and the motor is started with no load connected. Voltage readings are taken between T-1 and T-2. If the voltage is approximately 460, the markings are correct and may be permanently marked.

If the voltage reading is 400 V or less, interchange T-5 and T-7 *or* T-4 and T-9 and read the voltage again. If the voltage is approximately 230 V, interchange *both* T-5 with T-7 and T-4 with T-9. The readings should now be approximately 460 between leads T-1 and T-2. The leads connected together now are actually T-4 and T-7 and are marked permanently. The remaining lead in each group can now be marked T-9 and T-5 as indicated by Figure 9-7.

Connect one of the leads of the last coil group (not T-3) to T-9. If the reading is approximately 460 V between T-1 and T-3, the lead may be permanently marked T-6. If the reading is 400 V or less, interchange T-6 and T-8. A reading now of 460 V should exist between T-1 and T-3. T-6 is changed to T-8 and marked permanently and temporary T-8 is changed to T-6.

If all leads are now correctly marked, equal readings of approximately 460 V can be obtained between leads T-1, T-2, and T-3.

To double check the markings, the motor is shut off and reconnected using T-2, T-5, and T-7. T-2 is connected to the same line lead as T-1, lead T-5 is connected where T-4 was, and T-7 is hooked where T-9 was previously connected. When started, the motor should rotate the same direction as before.

Stop the motor and connect leads T-3, T-6, and T-8 to the line leads previously connected to T-2, T-5, and T-7, respectively, and when the motor is started it should still rotate in the same direction.

The motor is now ready for service and is connected in series for high or parallel for low voltage as indicated by the NEMA Standard connections shown in Figure 9-7.

Summary

The first step toward a reliable maintenance program is to prepare records. Obviously, records can take many forms and must be tailored to specific needs and resources. But, as a minimum, records on each motor should include:

- A complete description, including age and nameplate data.
- Location and application, keeping such notations up-to-date if motors are transferred to different areas or used for different purposes.
- Notations of scheduled preventive maintenance and previous repair work performed.
- Location of duplicate or interchangeable motors.
- An estimate of the motor's importance in the productive process to which it relates.

In determining which motors are likely to fail first, it is well to remember that motor failures generally are caused by either loading, age, vibration, contamination, or commutation problems.

Troubleshooting Charts

The troubleshooting chart (Figure 9-9) gives most motor problems found in the industry, along with their causes and remedies.

Symptoms	Probable Cause	Action or Items to Check
Hot bearings — general	Bent or sprung shaft.	Straighten or replace shaft.
	Excessive belt pull.	Decrease belt tension.
	Pulley too far away.	Move pulley closer to bearing.
	Pulley diameter too small.	Use larger pulleys.
	Misalignment.	Correct by realignment of drive.

9-9 Troubleshooting chart for electric motors.

Symptoms	Probable Cause	Action or Items to Check
Hot bearings — sleeve	Oil grooving in bearing obstructed by dirt.	Remove bracket or pedestal with bearing and clean oil grooves and bearing housing; renew oil.
	Bent or damaged oil rings.	Repair or replace oil rings.
	Oil too heavy or light.	Use a grade of oil recommended by motor manufacturer.

9-9 Troubleshooting chart for electric motors. *(Continued)*

Symptoms	Probable Cause	Action or Items to Check
Hot bearings — sleeve	Insufficient oil.	Fill reservoir to proper level in overflow plug with motor at rest.
	Too much end thrust.	Reduce thrust induced by driven machine or supply external means to carry thrust.
	Badly worn bearing.	Replace bearing.
Hot bearings — ball	Insufficient lubricant.	Replace bearing.

9-9 Troubleshooting chart for electric motors. *(Continued)*

Symptoms	Probable Cause	Action or Items to Check
Hot bearings — ball	Deterioration of grease or lubricant contaminated.	Remove old grease, wash bearings thoroughly in kerosene, and replace with new grease.
	Excess lubricant.	Reduce quantity of grease. Bearing should not be more than half filled.
	Heat from hot motor or external source.	Protect bearing by reducing motor temperature.

9-9 Troubleshooting chart for electric motors. *(Continued)*

Symptoms	Probable Cause	Action or Items to Check
Hot bearings — ball	Overloaded bearing.	Check alignment, side thrust, and end thrust.
	Broken ball or rough races.	Replace bearing; first clean housing thoroughly.
Oil leakage from overflow plugs	Stem of overflow plug not tight.	Remove, recement threads, replace, and tighten.

9-9 Troubleshooting chart for electric motors. *(Continued)*

Symptoms	Probable Cause	Action or Items to Check
Oil leakage from overflow plugs	Cracked or broken overflow plug.	Replace the plug.
	Plug cover not tight.	Requires cork gasket, or if screw type, may be tightened.
Motor dirty	Ventilation blocked, end windings filled with fine dust or lint.	Clean motor will run 10° to 30° C cooler. Dust may be cement, sawdust, rock dust, grain dust, coal dust, and the like. Dismantle entire motor and clean all windings and parts.

9-9 Troubleshooting chart for electric motors. *(Continued)*

Symptoms	Probable Cause	Action or Items to Check
Motor dirty	Rotor winding clogged.	Clean, grind, and undercut commutator. Clean and treat windings with good insulating varnish.
	Bearing and brackets coated inside.	Dust and wash with cleaning solvent.

9-9 Troubleshooting chart for electric motors. *(Continued)*

Symptoms	Probable Cause	Action or Items to Check
Motor wet	Subject to dripping.	Wipe motor and dry by circulating heated air through motor. Install drip or canopy type covers over motor for protection.
	Submerged in flood waters.	Dismantle and clean parts. Bake windings in oven at 105° C for 24 hours or until resistance to ground is sufficient. First make sure commutator bushing is drained of water.

9-9 Troubleshooting chart for electric motors. *(Continued)*

Symptoms	Probable Cause	Action or Items to Check
Fails to start	Circuit not complete.	Switch open, leads broken.
	Brushes not down on commutator.	Held up by brush springs, need replacement. Brushes worn out.
	Brushes stuck in holders.	Remove and sand, clean up brush boxes.
	Armature locked by frozen bearings in motor or main drive.	Remove brackets and replace bearings or recondition old bearings if inspection makes possible.

9-9 Troubleshooting chart for electric motors. *(Continued)*

Symptoms	Probable Cause	Action or Items to Check
Fails to start	Power may be off.	Check line connections to starter with light. Check contacts in starter.

9-9 Troubleshooting chart for electric motors. *(Continued)*

Symptoms	Probable Cause	Action or Items to Check
Motor starts, then stops and reverses direction of rotation	Shunt and series fields are bucking each other.	Reconnect either the shunt or series field so as to correct the polarity. Then connect armature leads for desired direction of rotation. The fields can be tried separately to determine the direction of rotation individually.
Motor does not come up to rated speed	Overload.	Check bearing to see if in first class condition with correct lubrication. Check driven load for excessive load or friction.

9-9 Troubleshooting chart for electric motors. *(Continued)*

Symptoms	Probable Cause	Action or Items to Check
Motor does not come up to rated speed	Starting resistance not all out.	Check starter to see if mechanically and electrically incorrect.
	Voltage low.	Measure voltage with meter and check with motor nameplate.
	Short circuit in armature windings or between bars.	For shorted armature inspect commutator for blackened bars and burned adjacent bars. Inspect windings for burned coils or wedges.

9-9 Troubleshooting chart for electric motors. *(Continued)*

Symptoms	Probable Cause	Action or Items to Check
Motor does not come up to rated speed	Starting heavy load with very weak field.	Check full field relay and possibilities of full field setting of the field rheostat.
	Motor off neutral.	Check for factory setting of brush rigging or test motor for true neutral setting.
	Motor cold.	Increase load on motor so as to increase its temperature, or add field rheostat to set speed.

9-9 Troubleshooting chart for electric motors. *(Continued)*

Symptoms	Probable Cause	Action or Items to Check
Motor runs too fast	Voltage above rated.	Correct voltage or get recommended change in air gap from manufacturer.
	Load too light.	Increase load or install fixed resistance in armature circuit.

9-9 Troubleshooting chart for electric motors. *(Continued)*

Symptoms	Probable Cause	Action or Items to Check
Motor runs too fast	Series coil reversed.	Reconnect coil leads in reverse.
	Series field coil shorted.	Install new or repaired coil.
	Neutral setting shifted off neutral.	Reset neutral by checking factory setting mark or testing for neutral.
	Part of shunt field rheostat or unnecessary resistance in field circuit.	Measure voltage across field and check with nameplate rating.

9-9 Troubleshooting chart for electric motors. *(Continued)*

Symptoms	Probable Cause	Action or Items to Check
Motor runs too fast	Motor ventilation restricted, causing hot shunt field.	Hot field is high in resistance; check causes for hot field in order to restore normal shunt field current.
Motor gaining speed steadily and increasing load does not slow it down	Unstable speed load regulation.	Inspect motor to see if off neutral. If series field has a shunt around the series circuit that can be removed, check series field to determine shorted turns.

9-9 Troubleshooting chart for electric motors. *(Continued)*

Symptoms	Probable Cause	Action or Items to Check
Motor gaining speed steadily and increasing load does not slow it down	Reversed field coil shunt or series.	Test with compass and reconnect coil.
	Too strong a commutating pole or commutating pole air gap too small.	Check with factory for recommended change in coils or air gap.
Motor runs too slow continuously	Voltage below rated.	Check voltage and correct to value on nameplate.
	Motor operates cold.	Motor may run 20% slow due to light load. Install smaller motor.

9-9 Troubleshooting chart for electric motors. *(Continued)*

Symptoms	Probable Cause	Action or Items to Check
Motor runs too slow continuously	Overload.	Check bearings of motors and the drive to see if in first class condition. Check for excessive friction in drive.
	Neutral setting shifted.	Check for factory setting of brush rigging or test for true neutral setting.
	Armature has shorted coils or commutator bars.	Remove armature to repair shop and put in first class condition.

9-9 Troubleshooting chart for electric motors. *(Continued)*

Symptoms	Probable Cause	Action or Items to Check
Motor overheats or runs hot	Overloaded and draws 25% to 50% more current than rated.	Reduce load by reducing speed or gearing in the drive or loading in the drive.
	Voltage above rated.	Motor runs drive above rated speed requiring excessive hp. Reduce voltage to nameplate rating.
	Inadequately ventilated.	Location of motor should be changed.

9-9 Troubleshooting chart for electric motors. *(Continued)*

Symptoms	Probable Cause	Action or Items to Check
Motor overheats or runs hot	Draws excessive current due to shorted coil.	Repair armature coils or install new coil.
	Grounds in armature such as two grounds which constitute a short.	Locate grounds and repair or rewind with new set of coils.
	Armature rubs pole faces due to off-center rotor causing friction and excessive current.	Check brackets or pedestals to center rotor and determine condition of bearing wear for bearing replacement.

9-9 Troubleshooting chart for electric motors. *(Continued)*

Symptoms	Probable Cause	Action or Items to Check
Hot armature	Core hot in one spot indicating shorted punching and high iron loss.	Sometimes full slot metal wedges have been used for balancing. These should be removed and other means of balancing investigated.
	Brush tension too high.	Limit pressure to 2 to 2½ psi. Check brush density and limit to density recommended by the brush manufacturer.

9-9 Troubleshooting chart for electric motors. *(Continued)*

Symptoms	Probable Cause	Action or Items to Check
Hot armature	Punching uninsulated. Punching has been turned or band grooves machined in the core. Machined slots.	No-load running of motor will indicate hot core and drawing high no-load armature current. Replace core and rewind armature. If necessary to add band grooves grind into core. Check temperature on core with thermometer: not to exceed 90°C.
Hot commutator	Brushes off neutral.	Reset neutral.

9-9 Troubleshooting chart for electric motors. *(Continued)*

Symptoms	Probable Cause	Action or Items to Check
Hot commutator	Brush grade too abrasive.	Get recommendation from manufacturer.
	Shorted bars.	Investigate commutator mica and undercutting, and repair.

9-9 Troubleshooting chart for electric motors. *(Continued)*

Symptoms	Probable Cause	Action or Items to Check
Hot commutator	Inadequate ventilation.	Check as for hot motor.
Hot fields	Voltage too high.	Check with meter and thermometer and correct voltage to nameplate value.
	Shorted turns or grounded turns.	Repair or replace with new coil.
	Resistance of each coil not the same.	Check each individual coil for equal resistance within 10% and if one coil is too low replace coil.

9-9 Troubleshooting chart for electric motors. *(Continued)*

Symptoms	Probable Cause	Action or Items to Check
Hot fields	Inadequate ventilation.	Check as for hot motor.
	Coil not large enough to radiate its loss wattage.	Coils should be replaced if room is available in motor.
Motor vibrates, indicating unbalance	Armature out of balance.	Remove and statically balance or balance in dynamic balancing machine.
	Misalignment.	Realign.

9-9 Troubleshooting chart for electric motors. *(Continued)*

Symptoms	Probable Cause	Action or Items to Check
Motor vibrates, indicating unbalance	Loose or eccentric pulley.	Tighten pulley on shaft or correct eccentric pulley.
	Belt or chain whip.	Adjust belt tension.
	Mismating of gear and pinion.	Recut, realign, or replace parts.
	Unbalance in coupling.	Rebalance coupling.
	Bent shaft.	Replace or straighten shaft.

9-9 Troubleshooting chart for electric motors. *(Continued)*

Symptoms	Probable Cause	Action or Items to Check
Motor vibrates, indicating unbalance	Foundation inadequate.	Stiffen mounting place members.
	Motor loosely mounted.	Tighten holding-down bolts.
	Motor feet uneven.	Add shims under foot pads to mount each foot tight.
Motor sparks at brushes or does not commutate	Neutral setting not true neutral.	Check and set on factory setting or test for true neutral.

9-9 Troubleshooting chart for electric motors. *(Continued)*

Symptoms	Probable Cause	Action or Items to Check
Motor sparks at brushes or does not commutate	Commutator rough.	Grind and roll edge of each bar.
	Commutator eccentric.	Turn, grind, and roll, commutator.
	Mica high—not undercut.	Undercut mica.
	Commutating pole strength too great causing overcompensation or strength too weak indicating under compensation.	Check with manufacturer for correct change in air gap or new coils for the commutating coils.

9-9 Troubleshooting chart for electric motors. *(Continued)*

Symptoms	Probable Cause	Action or Items to Check
Motor sparks at brushes or does not commutate	Shorted commutating pole turns.	Repair coils or install new coils.
	Shorted armature coils on commutator bars.	Repair armature.
	Open-circuited coils.	Repair armature.
	Poor soldered connection to commutator bars.	Resolder with proper alloy of tin solder.

9-9 Troubleshooting chart for electric motors. *(Continued)*

Symptoms	Probable Cause	Action or Items to Check
Motor sparks at brushes or does not commutate	High bar or loose bar in commutator at high speeds.	Inspect commutator nut or bolts and retighten and grind commutator face.
	Brush grade wrong type. Brush pressure too light, current density excessive, brushes stuck in holders. Brush shunts loose.	Inspect brushes and replace as necessary.
	Brushes chatter due to dirty film on commutator.	Resurface commutator face and check for change in brushes.

9-9 Troubleshooting chart for electric motors. *(Continued)*

Symptoms	Probable Cause	Action or Items to Check
Motor sparks at brushes or does not commutate	Vibration.	Eliminate cause of vibration by checking mounting and balance of rotor.
Brush wear excessive	Brushes too soft.	Blow dust from motor and replace brushes with grade recommended by manufacturer.
	Cummutator rough.	Grind commutator face.

9-9 Troubleshooting chart for electric motors. *(Continued)*

Symptoms	Probable Cause	Action or Items to Check
Brush wear excessive	Off neutral setting.	Recheck factory neutral or test for true neutral.
	Abrasive dust in ventilating air.	Reface brushes and correct condition by protecting motor.

9-9 Troubleshooting chart for electric motors. *(Continued)*

Symptoms	Probable Cause	Action or Items to Check
Brush wear excessive	Brush tension excessive.	Adjust spring pressure not to exceed 2 to 2½ psi.
	Electrical wear due to loss of film on commutator face.	Resurface brush faces and commutator face. Check for change in brush grade.
	Threading and grooving.	Resurface brush faces and commutator face. Check for change in brush grade.
	Oil or grease from atmosphere or bearings.	Correct oil condition and surface brush faces and commutator.

9-9 Troubleshooting chart for electric motors. *(Continued)*

Symptoms	Probable Cause	Action or Items to Check
Motor noisy	Brush singing.	Check brush angle and commutator coating, resurface commutator.
	Brush chatter.	Resurface commutator and brush face.
	Motor loosely mounted.	Tighten foundation bolts.

9-9 Troubleshooting chart for electric motors. *(Continued)*

Symptoms	Probable Cause	Action or Items to Check
Motor noisy	Foundation hollow and acts as sounding board.	Coat underside with sound-proofing material.
	Strained frame.	Shim motor feet for equal mounting.
	Armature punching loose.	Replace core on armature.
	Armature rubs pole faces.	Recenter by replacing bearings or relocating brackets or pedestals.
	Magnetic hum.	Refer to manufacturer.

9-9 Troubleshooting chart for electric motors. *(Continued)*

Symptoms	Probable Cause	Action or Items to Check
Motor noisy	Belt slap or pounding.	Check condition of belt and adjust belt tension.
	Excessive current load.	May not cause overheating but check chart for correction of shorted or grounded coils.
	Mechanical vibration.	Check chart for causes of vibration.
	Noisy bearings.	Check alignment, loading of bearings, and lubrication.

9-9 Troubleshooting chart for electric motors. *(Continued)*

Symptoms	Probable Cause	Action or Items to Check
Motor stalls	Wrong application.	Change type or size. Consult manufacturer.
	Overloaded motor.	Reduce load.
	Low motor voltage.	See that nameplate voltage is maintained.
	Open circuit.	Fuses blown, check overload relay, starter, and pushbuttons.

9-9 Troubleshooting chart for electric motors. *(Continued)*

Symptoms	Probable Cause	Action or Items to Check
Motor stalls	Incorrect control resistance of wound rotor.	Check control sequence. Replace broken resistors. Repair open circuits.
Motor connected but does not start	One phase open.	Reconnect open phase.
	Motor may be overloaded.	Reduce load.
	Rotor defective.	Look for broken bars or rings.
	Poor stator coil connection.	Remove end bells.

9-9 Troubleshooting chart for electric motors. *(Continued)*

Symptoms	Probable Cause	Action or Items to Check
Motor runs and then dies down	Power failure.	Check for loose connections to line, to fuses, and to control.
Motor does not come up to speed	Not applied properly.	Consult supplier for proper type.
	Voltage too low at motor terminals because of line drop.	Use higher voltage on transformer terminals or reduce load.

9-9 Troubleshooting chart for electric motors. *(Continued)*

Symptoms	Probable Cause	Action or Items to Check
Motor does not come up to speed	If wound rotor, improper control operation of secondary resistance.	Correct secondary control.
	Starting load too high.	Check load motor is supposed to carry at start.
	Low pull-in torque of synchronous motor.	Change rotor starting resistance or change rotor design.
	Check that all brushes are riding on rings.	Check secondary connections.

9-9 Troubleshooting chart for electric motors. *(Continued)*

Symptoms	Probable Cause	Action or Items to Check
Motor does not come up to speed	Broken rotor bars.	Look for cracks near the rings. A new rotor may be required as repairs are usually temporary.
	Open primary circuit.	Locate fault with testing device and repair.
Motor takes too long to accelerate	Excess loading.	Reduce load.
	Poor circuit.	Check for excessive voltage drop.

9-9 Troubleshooting chart for electric motors. *(Continued)*

Symptoms	Probable Cause	Action or Items to Check
Motor takes too long to accelerate	Defective squirrel-cage rotor.	Replace with new rotor.
	Applied voltage too low.	Get power company to increase voltage tap.
Wrong rotation	Wrong sequence of phase.	Reverse connections of motor or at switchboard.
Motor overheats while running under load	Check for overload.	Reduce load.

9-9 Troubleshooting chart for electric motors. *(Continued)*

Symptoms	Probable Cause	Action or Items to Check
Motor overheats while running under load	Wrong blowers or air shields, may be clogged with dirt and prevent proper ventilation of motor.	Good ventilation is manifest when a continuous stream of air leaves the motor. If not, check manufacturer.
	Motor may have one phase open.	Check to make sure that all leads are well connected.
	Grounded coil.	Locate and repair.
	Unbalanced terminal voltage.	Check for faulty leads, connections, and transformers.

9-9 Troubleshooting chart for electric motors. *(Continued)*

Symptoms	Probable Cause	Action or Items to Check
Motor overheats while running under load	Shorted stator coil.	Repair and then check wattmeter reading.
	Voltage too high or low.	Check terminals of motor with voltmeter.
	Rotor rubs stator bore.	If not poor machining, replace worn bearings.

9-9 Troubleshooting chart for electric motors. *(Continued)*

Symptoms	Probable Cause	Action or Items to Check
Motor vibrates after corrections have been made	Motor misaligned.	Realign.
	Weak foundations.	Strengthen base.
	Coupling out of balance.	Balance coupling.
	Driven equipment unbalanced.	Rebalance driven equipment.
	Defective ball bearing.	Replace bearing.
	Bearings not in line.	Line up properly.

9-9 Troubleshooting chart for electric motors. *(Continued)*

Symptoms	Probable Cause	Action or Items to Check
Motor vibrates afer corrections have been made	Balancing weights shifted.	Rebalance rotor.
	Wound rotor coils replaced.	Rebalance rotor.
	Polyphase motor running single phase.	Check for open circuit.
	Excessive end play.	Adjust bearing or add washer.

9-9 Troubleshooting chart for electric motors. *(Continued)*

Symptoms	Probable Cause	Action or Items to Check
Unbalanced line current on polyphase motors during normal operation	Unequal terminal voltages.	Check leads and connections.
	Single-phase operation.	Check for open contacts.
	Poor rotor contacts in control wound-rotor resistance.	Check control devices.
	Brushes not in proper position in wound rotor.	See that brushes are properly seated and shunts in good condition.

9-9 Troubleshooting chart for electric motors. *(Continued)*

Symptoms	Probable Cause	Action or Items to Check
Scraping noise	Fan rubbing air shield.	Remove interference.
	Fan striking insulation.	Clear fan.
	Loose on bedplate.	Tighten holding bolts.
Magnetic noise	Air gap not uniform.	Check and correct bracket fits or bearing.
	Loose bearings.	Correct or renew.
	Rotor unbalance.	Rebalance.

9-9 Troubleshooting chart for electric motors. *(Continued)*

CHAPTER 10

Troubleshooting Motor Bearings

Alternating-current motors account for a high percentage of electrical repair work. A high proportion of these failures can be traced to faulty bearings. Consequently, most industrial establishments place heavy emphasis on the proper handling, repair, and maintenance of types of sleeve and ball bearings.

Bearing failure can occur in newer motors with high-quality bearings as frequently as in older motors equipped with less reliable bearings. A notable exception is motors equipped with sealed bearings which are much less prone to failure.

Types of Bearings

There are many types of bearings, but ball bearings seem to be the most common. This type of bearing is found on various-size motors and their construction may be:

- Open
- Single shielded
- Double shielded

- Sealed
- Double row and other special types

Open bearings, as the name implies, are open construction and must be installed in a sealed housing. These bearings are less apt to cause churning of grease, and are therefore used mostly on large motors.

The single-shield bearing has a shield on one side to preclude grease from the motor windings. Doubleshielded bearings have a shield on both sides of the bearing. This type of bearing is less susceptible to contamination and, because of its design, reduces the possibility of over-greasing. Sealed bearings have, on each side of the bearing, double shields which form an excellent seal. This bearing requires no maintenance, affords protection from contamination at all times, and does not require regreasing. It is normally used on small or medium-size motors.

The largest motors usually are furnished with oil-ring sleeve bearings. And some of the fractional-horsepower motors are equipped with plain sleeve bearings.

Each bearing type has characteristics which make it the best choice for a certain application. Replacement should be made with the same type bearings. The following list of functions provide a basic understanding of bearing application, a guide to analysis of bearing troubles due to misapplication, and emphasize the importance of proper replacement.

Figure 10-1 shows several types of bearings used in electric motors. The following is a brief description of each:

Self-aligning ball bearing

 Spherical-roller bearing

 Single-row, deep-groove ball bearing

 Cylindrical-roller bearing

 Angular-contact ball bearing

 Ball-thrust bearing

 Spherical-roller thrust bearing

 Double-row, deep-groove ball bearing

 Tapered-roller bearing

10-1 Various bearing types.

Self-aligning ball bearings: The self-aligning ball bearing, with two rows of balls rolling on the spherical surface of the outer ring, compensates for angular misalignment resulting from errors in mounting, shaft deflection, and distortion of the foundation. It is impossible for this bearing to exert any bending influence on the shaft—a most important consideration in applications requiring extreme accuracy, at high speeds. Self-aligning ball bearings are used for radial loads and moderate thrust loads in either direction.

Single-row, deep-groove ball bearings: The single-row, deep-groove ball bearing will sustain, in addition to radial load, a substantial thrust load in either direction, even at very high speeds. This advantage results from the intimate contact existing between the balls and the deep, continuous groove in each ring. When using this type of bearing, careful alignment between the shaft and housing is essential. This bearing is also available with seals and shields, which serves to exclude dirt and retain lubricant.

Angular-contact ball bearings: The angular-contact ball bearing supports a heavy thrust load in one direction, sometimes combined with a moderate radial load. A steep contact angle, assuring the highest thrust capacity and axial rigidity, is obtained by a high thrust-supporting shoulder on the inner ring and a similar high shoulder on the opposite side of the outer ring. These bearings can be mounted singly or, when the sides are flush ground, in tandem for constant thrust in one direction; mounted in pairs, also when sides are flush ground, for a combined load, either face-to-face or back-to-back.

Double-row, deep-groove ball bearings: The double-row, deep-groove ball bearing embodies the same principle of design as the single-row bearing. However, this bearing has a lower axial displacement than occurs in the single-row design, substantial thrust capacity in either direction, and high radial capacity due to the two rows of balls.

Self-aligning ball bearing

 Spherical-roller bearing

 Single-row, deep-groove ball bearing

 Cylindrical-roller bearing

 Angular-contact ball bearing

 Ball-thrust bearing

 Spherical-roller thrust bearing

 Double-row, deep-groove ball bearing

 Tapered-roller bearing

10-1 Various bearing types.

Self-aligning ball bearings: The self-aligning ball bearing, with two rows of balls rolling on the spherical surface of the outer ring, compensates for angular misalignment resulting from errors in mounting, shaft deflection, and distortion of the foundation. It is impossible for this bearing to exert any bending influence on the shaft—a most important consideration in applications requiring extreme accuracy, at high speeds. Self-aligning ball bearings are used for radial loads and moderate thrust loads in either direction.

Single-row, deep-groove ball bearings: The single-row, deep-groove ball bearing will sustain, in addition to radial load, a substantial thrust load in either direction, even at very high speeds. This advantage results from the intimate contact existing between the balls and the deep, continuous groove in each ring. When using this type of bearing, careful alignment between the shaft and housing is essential. This bearing is also available with seals and shields, which serves to exclude dirt and retain lubricant.

Angular-contact ball bearings: The angular-contact ball bearing supports a heavy thrust load in one direction, sometimes combined with a moderate radial load. A steep contact angle, assuring the highest thrust capacity and axial rigidity, is obtained by a high thrust-supporting shoulder on the inner ring and a similar high shoulder on the opposite side of the outer ring. These bearings can be mounted singly or, when the sides are flush ground, in tandem for constant thrust in one direction; mounted in pairs, also when sides are flush ground, for a combined load, either face-to-face or back-to-back.

Double-row, deep-groove ball bearings: The double-row, deep-groove ball bearing embodies the same principle of design as the single-row bearing. However, this bearing has a lower axial displacement than occurs in the single-row design, substantial thrust capacity in either direction, and high radial capacity due to the two rows of balls.

DO's

DO work with clean tools, in clean surroundings.

DO remove all outside dirt from housing before exposing bearing.

DO treat a used bearing as carefully as a new one.

DO use clean solvents and flushing oils.

DO lay bearings out on clean paper or cloth.

DO protect disassembled bearings from dirt and moisture.

DO use clean, lint-free rags to wipe bearings.

DO keep bearings wrapped in oil-proof paper when not in use.

DO clean outside of housing before replacing bearings.

DO keep bearing lubricants clean when applying and cover containers when not in use.

DO be sure shaft size is within specified tolerances recommended for the bearing.

DO store bearings in original unopened cartons in a dry place.

DO use a clean, short-bristle brush with firmly embedded bristles to remove dirt, scale, or chips.

10-2 Do's and don'ts for ball-bearing assembly, maintenance, and lubrication.

DO's *(Cont.)*

DO be certain that, when installed, the bearing is square with and held firmly against the shaft shoulder.

DO follow lubricating instructions supplied with the machinery. Use only grease where grease is specified; use only oil where oil is specified. Be sure to use the exact kind of lubricant called for.

DO handle grease with clean paddles or grease runs. Store grease in clean containers. Keep grease containers covered.

DON'Ts

DON'T work under the handicap of poor tools, rough benches, or dirty surroundings.

DON'T use dirty, brittle, or chipped tools.

DON'T handle bearings with dirty, moist hands.

DON'T spin uncleaned bearings.

DON'T spin any bearings with compressed air.

DON'T use same container for cleaning and final rinse of bearings.

DON'T scratch or nick bearing surfaces.

10-2 Do's and don'ts for ball-bearing assembly, maintenance, and lubrication. *(Continued)*

DON'TS *(Cont.)*

DON'T remove grease or oil from new bearings.

DON'T use incorrect kind or amount of lubricant.

DON'T use a bearing as a gauge to check either the housing bore or the shaft fit.

DON'T install a bearing on a shaft that shows excessive wear.

DON'T open a carton until the bearing is ready for installation.

DON'T judge the condition of a bearing until after it has been cleaned.

DON'T pound directly on a bearing or ring, when installing, as this may cause damage to shaft and bearing.

DON'T overfill when lubricating. Excess grease and oil will ooze out of the overfilled housings past seals and closures, collect dirt, and cause trouble. Too much lubricant will also cause overheating, particularly where bearings operate at high speeds.

DON'T permit any machine to stand inoperative for months without turning it over periodically. This prevents moisture which may condense in a standing bearing from causing corrosion.

10-2 Do's and don'ts for ball-bearing assembly, maintenance, and lubrication. *(Continued)*

Lubrication Procedure

For effective motor lubrication, cleanliness and use of the proper lubricant are of paramount importance.

When greasing a ball-bearing motor, the bearing housing, grease gun, and fittings are wiped clean. Great care must be taken to keep dirt out of the bearing when greasing. Next, the relief plug is removed from the bottom of the bearing housing. This is done to prevent excessive pressure from building up inside the bearing housing during greasing. Grease is then added, with the motor running if possible, until it begins to flow from the relief hole. Allow the motor to run from 5 to 10 minutes to expel excess grease. Then the relief plug is replaced and the bearing housing is cleaned.

It is important to avoid over-greasing. When too much grease is forced into a bearing, a churning of the lubricant occurs, resulting in high temperature and eventual bearing failure.

On motors that do not have a relief hole, grease should be applied sparingly. If possible, disassemble the motor and repack the bearing housing with the proper amount of grease. During this procedure, always maintain strict cleanliness. Contamination and overgreasing of bearings are the major causes of bearing failure.

For sleeve bearings, use only the recommended oil for particular service conditions. Observing careful cleanliness, old oil is removed and new oil is added until the oil level reaches the "full" line on the oil sight gauge. This is done only when the motor is not running.

Testing Bearings

Two of the most effective tests are what might be called the "feel" test and the "sound" test. When performing the "feel" test, if, while the motor is running, the bearing housing feels overly hot to the touch, it is probably malfunctioning.

> **Note**
> Some bearings may operate safely up to about 85°C

During the "sound" test, listen for foreign noises coming from the motor. Also, one end of a steel rod (about 3 feet long and ½ inch in diameter) may be placed on the bearing housing while the other end is held against the ear. The rod acts as an amplifier, transmitting unusual sounds such as thumping or grinding, which would indicate a failing bearing. Special listening devices, such as the transistorized stethoscope, can also be used for the purpose.

Additional checks are usually in the form of checking the air gap on sleeve-bearing motors periodically. These tests, performed with a feeler gauge, indicate when a bearing begins to wear. Four measurements should be taken about 90 degrees apart around the rotor periphery. These measurements are recorded and compared with earlier readings, providing a check on the condition of bearings.

Motors should also be checked for end play. Ballbearing motors should have about ½₃₂ inch to ⅟₁₆ inch end play. Sleeve-bearing motors may have up to ½ inch end play.

On large sleeve bearings the oil level should be checked periodically, and the oil is visually inspected for contamination. If it is possible, the oil rings should be checked when the motor is operating.

Other inspections include checking for misaligned or bent shafts and for excessive belt pressure.

The troubleshooting chart in Figure 10-3 lists the most common problems with motor bearings.

Symptoms	Probable Cause	Action or Items to Check
Hot Bearings — General	Bent or sprung shaft.	Straighten or replace shaft.
	Excessive belt pull.	Decrease belt tension.
	Pulley too far away.	Move pulley closer to bearing.
	Pulley diameter too small.	Use larger pulleys.
	Misalignment.	Correct by realignment of drive.

10-3 Troubleshooting chart for motor bearings.

Symptoms	Probable Cause	Action or Items to Check
Hot Bearings — Sleeve	Oil grooving in bearing obstructed by dirt.	Remove bracket or pedestal with bearing and clean oil grooves and bearing housing; renew oil.
	Bent or damaged oil rings.	Repair or replace oil rings.
	Oil too heavy.	Use a recommended lighter oil.
	Oil too light.	Use a recommended heavier oil.

10-3 Troubleshooting chart for motor bearings. *(Continued)*

Symptoms	Probable Cause	Action or Items to Check
Hot Bearings– Sleeve	Insufficient oil.	Fill reservoir to proper level in overflow plug with motor at rest.
	Too much end thrust.	Reduce thrust induced by driven machine or supply external means to carry thrust.
	Badly worn bearing.	Replace bearing.

10-3 Troubleshooting chart for motor bearings. *(Continued)*

Symptoms	Probable Cause	Action or Items to Check
Hot Bearings — Ball	Insufficient grease.	Replace bearing.
	Deterioration of grease or lubricant contaminated.	Remove old grease, wash bearings thoroughly in kerosene, and replace with new grease.
	Excess lubricant.	Reduce quantity of grease. Bearing should not be more than half filled.

10-3 Troubleshooting chart for motor bearings. *(Continued)*

Symptoms	Probable Cause	Action or Items to Check
Hot Bearings — Ball	Heat from hot motor or external source.	Protect bearing by reducing motor temperature.
	Overloaded bearing.	Check alignment, side thrust, and end thrust.
	Broken ball or rough races.	Replace bearing; first clean housing thoroughly.

10-3 Troubleshooting chart for motor bearings. *(Continued)*

CHAPTER 11

Troubleshooting Relays and Contactors

A relay is an electromagnetic or solid-state device used in control circuits of magnetic motor starters, heaters, solenoids, timers, and other devices. Relays are generally used to amplify the contact capability or multiply the switching functions of a control device, and for remote control applications. Relays are manufactured in a large number of different configurations; Figure 11-1 shows a type of relay often used to control small, single-phase motors and other light loads such as heaters or pilot lights. Magnetic mechanical relays are still the most common, although solid-state relays are finding their way into all types of control applications.

Contactors are a type of electromagnetic apparatus similar in construction and operation to relays, but built to handle much higher currents (Figure 11-2) involved in applications such as switching large banks of stadium lights on and off.

11-1 Single-pole, single-throw (SPST) relay rated 30 amperes, 600 volts. *(Courtesy Square D Company.)*

11-2 NEMA size 1 contactor rated 10HP, 575 volts. *(Courtesy Square D Company.)*

Symptoms	Probable Cause	Action or Items to Check
Failure to pull in	Either no voltage or low voltage at coil terminals.	Blown fuse, open line switch, or break in wiring.
		Line voltage below normal.
		Overload relay open or set too low.
		Tripping toggle (nonautomatic breakers) fouled.
		Control level or start button in OFF position.

11-3 Contactor and relay troubleshooting chart.

Symptoms	Probable Cause	Action or Items to Check
Failure to pull in	Either no voltage or low voltage at coil terminals	Pull-in circuit open, shorted, or grounded
		Contacts in protective or controlling circuit open or one of their pigtail connections broken
	Operating coil open or grounded	Inspect and test coil
	Loose or disconnected coil lead wire	Inspect and correct

11-3 Contactor and relay troubleshooting chart. *(Continued)*

Symptoms	Probable Cause	Action or Items to Check
Failure to pull in	Excessive magnet gap, improper alignment	Inspect and correct
	Armature obstructed or gumming deposits between armature and pole face	Inspect and correct
	Binding caused by deformed or gummy hinge	Replace if bent; degrease if gummed up
	Excessive armature spring force	Lessen spring force

11-3 Contactor and relay troubleshooting chart. *(Continued)*

Symptoms	Probable Cause	Action or Items to Check
Failure to pull in	Normally closed contacts welded together	Replace contacts
Failure of equipment to start with contactor closed	One contact not closing	Replace set of contacts
	Contacts burned	Use emery cloth to lightly burnish or replace
	Contact pigtail connection broken	Replace

11-3 Contactor and relay troubleshooting chart. *(Continued)*

Symptoms	Probable Cause	Action or Items to Check
Failure to drop out	Operating coil is energized	Contacts in controlling or protective tripping circuits closed, shorted, or shunted
		Tripping devices defective; such as overload tripping toggles do not strike release, undervoltage relay plunger stuck or out of adjustment, defective stop button, or defective time-delay escape mechanism (closed air vent)

11-3 Contactor and relay troubleshooting chart. *(Continued)*

Symptoms	Probable Cause	Action or Items to Check
Failure to drop out	Operating coil is energized	Current supplied over an unintentional path, due to grounds, defective insulation, pencil markings, moisture, or lacquer chipped off relay's base
	Residual magnetism excessive, due to armature closed tightly against pole face	Adjust or replace

11-3 Contactor and relay troubleshooting chart. *(Continued)*

Symptoms	Probable Cause	Action or Items to Check
Failure to drop out	Armature obstructed or gumming deposits between armature and pole face	Clear and clean
	Binding caused by deformed or gummy hinge	Replace or degrease
	Contact pressure spring or armature spring too weak or improperly adjusted	Replace or adjust
	Improper mounting position (upside down)	Mount correctly

11-3 Contactor and relay troubleshooting chart. *(Continued)*

Symptoms	Probable Cause	Action or Items to Check
Failure to drop out	Normally opened contacts welded together	Replace contacts
Time delay relays operate too fast	Escapement mechanism faulty	Air escapes too freely due to holes in bellows or open air vent
		Worn dashpot plunger, or dashpot oil too thin
	Nonmagnetic shim in air gap too large or armature spring too strong	Adjust or replace
	Magnets out of adjustment	Adjust

11-3 Contactor and relay troubleshooting chart. *(Continued)*

Symptoms	Probable Cause	Action or Items to Check
Contacts pitted or discolored	Contacts overheated from overload	Reduce load and replace contacts
	Contacts not fitted properly	Refit
	Barriers broken from rough usage or breakers closing with too much force	Replace barriers; check for high voltage
	Wiping action of contacts on closing is insufficient	Correct
	Excessive chatter or hum	Check shaded poles

11-3 Contactor and relay troubleshooting chart. *(Continued)*

Symptoms	Probable Cause	Action or Items to Check
Contacts pitted or discolored	Exposure to weather, dripping water, salt air, or vibration	Use suitable NEMA enclosure type.
Excessive chatter or hum	Vibration of surrounding devices communicated to relay	Arrange differently
	Relay receiving contradictory signals	Correct
	Bouncing of controlling or protective contacts	Correct

11-3 Contactor and relay troubleshooting chart. *(Continued)*

Symptoms	Probable Cause	Action or Items to Check
Excessive chatter or hum	Line voltage fluctuating or insufficient	Consider using buck-and-boost transformers
	Excessive coil circuit resistance	Lessen resistance
	Armature spring or contact pressure spring too strong	Lessen
	Excessive coil voltage drop on closing	Correct
	Improper antifreeze pin location	Locate in correct position

11-3 Contactor and relay troubleshooting chart. *(Continued)*

Symptoms	Probable Cause	Action or Items to Check
Excessive chatter or hum	Free movement of armature hindered due to deformed parts or dirt	Clean and replace appropriate parts
Excessive coil temperature	Excessive current or voltage	Reduce load or adjust taps on transformer
	Short circuit in coil	Replace coil
	Excessive eddy current and hysteresis	Correct
	High room temperature	Replace with properly rated relay

11-3 Contactor and relay troubleshooting chart. *(Continued)*

CHAPTER 12

Troubleshooting Power Quality Problems

The service panelboard (Figure 12-1) is a convenient place to take the pulse of an electrical system. Power quality problems can often be diagnosed at the panel by following this step-by-step approach:

- Voltage levels (steady state) and voltage stability (surges and sags)
- Current balance (phase loading)
- Harmonics
- Grounding
- Overheated terminals and connections
- Faulty or marginal circuit breakers

Voltage Levels and Stability

Voltage levels
Check voltage levels at the main panel terminals and each branch circuit. Voltage at the panel should ideally

12-1 Main distribution panelboard. *(Courtesy Square D Company.)*

be 120/208 or 277/480 volts, three-phase, four-wire. Voltage at receptacles or utilization equipment may be lower due to voltage drop on branch circuits, but should ideally be no less than 115/200 or 265/460 volts. For safety, take voltage measurements on the load side of main or branch circuit breakers whenever possible. This precaution helps protect the test instrument and operator from potential fault currents on feeders (see Figure 12-2).

Low voltage causes electric motors to run slower than their design speed, incandescent lights to burn

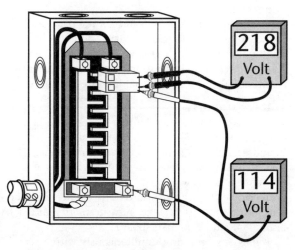

12-2 Safe voltage measurement technique at panelboard.

dimmer, starting problems for fluorescent and HID lamps, and performance problems for electronic and data devices. Overvoltage causes motors to run faster, shortens incandescent lamp life, and can damage sensitive electronic components.

Most electrical and electronic equipment is designed to tolerate a range of plus or minus 10 percent of rated voltage and still operate satisfactorily. However, panelboard voltages in the range of 115/200 or 265/460 volts will probably translate into unacceptably low voltages at receptacles or utilization equipment, by the time voltage drop on the branch circuit conductors is taken into account.

Common causes of low voltage at the panel are low tap settings at transformers, feeder conductors that are too long or too small, and loose connections. The first condition results in lower supply voltage; the latter two result in higher impedance that increases voltage drop.

Voltage stability

Voltage sags can be caused either by loads on branch circuits, or elsewhere in the distribution system, including utility-generated sags and brownouts. This is most easily analyzed using a dual-channel instrument such as a power quality analyzer or hand-held oscilloscope that can measure both voltage and current simultaneously. Take measurements at each branch circuit in the panelboard.

Voltage sag occurring simultaneously with a *current surge* usually indicates a problem downstream of the measurement point. This would be a load-related disturbance on the branch circuit.

Voltage sag occurring simultaneously with a *current sag* usually indicates a problem upstream of the measurement point, originating elsewhere in the distribution system. Typical source-related disturbances include large three-phase motors coming on line (starting) or sags in the utility network.

Current Loading

Measure the current on each feeder phase and branch circuit (Figure 12-3). It is important to make these measurements using a true-RMS clamp-on ammeter or digital multimeter (DMM). Because the combination of fundamental and harmonic currents results in a distorted waveform, a lower-cost average-sensing meter will tend to read low, leading you to assume that circuits are more lightly loaded than they actually are.

Loading on the three phases should be as balanced as possible. Unbalanced current will return on the neutral conductor, which may already be carrying a high load due to harmonics caused by nonlinear loads. In an ideal, balanced, three-phase electrical distribution system, there is little or no load on the neutral.

Neither the panel feeder nor branch circuits should be loaded to the maximum allowable limit (80 percent of the overcurrent device rating, for continuous loads). There should be some spare capacity to allow for harmonic currents.

Harmonics

Nonlinear single-phase loads can cause third harmonics (also called *triplen* harmonics) to add up in

12-3 Branch-circuit panelboard. *(Courtesy Square D Company.)*

the neutrals of the three-phase power systems, as shown in Figure 12-4. Nonlinear loads include such common electrical equipment as switched-mode power supplies used in computers and their peripherals, and fluorescent or HID fixture ballasts. Overloaded neutrals are a potential fire hazard because, unlike phase conductors, they are not protected by an overcurrent device.

Third harmonics can overload system neutral conductors even where care has been taken to balance utilization equipment loading among the three phases. For this reason the National Electrical Code requires that "where the major portion of the load consists of nonlinear loads...the neutral shall be considered a current-carrying conductor" (Article 310, Note 10c to Ampacity Tables of 0 to 2000 Volts).

In effect, this requires that neutral conductors of such three-phase, four-wire systems be at least the same size as the phase conductors. In practice, neutrals of systems serving a high proportion of nonlinear loads (such as office areas with multiple computers and fluorescent lighting) are sometimes even larger, up to double the size of the associated phase conductors (Figure 12-4).

Multiwire branch circuits

Common neutrals shared by either two or three single-phase branch circuits are subject to the same overloading, due to asymmetrical loading and third harmonics, as neutrals of three-phase panel feeders.

Harmonic currents in feeder or branch circuit neutral conductors can be measured using a clamp-on ammeter or DMM, or by using a probe-type meter to

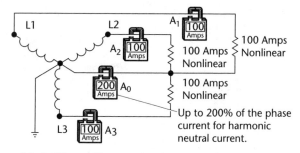

12-4 Effect of harmonics due to nonlinear loads.

measure the potential from neutral to ground (see Figure 12-5).

Grounding

The neutral and grounding electrode conductor should be bonded together only once, at the service entrance or distribution point of a separately derived system. Other neutral-ground connections elsewhere in the system, such as subpanels or receptacle outlets, are a violation of the National Electrical Code. Unfortunately, such "illegal" downstream neutral-ground connections are also very common, and they are frequently a source of power quality problems.

When the neutral and grounding electrode conductors are bonded at a subpanel or other location, the ground path becomes a parallel return path for normal load current, and there will be measurable current on the ground.

To determine whether improper connections exist, measure the neutral current and then the current on the grounding electrode (green) conductor and look at the ratio between them. For example, if the neutral current is 70 amperes and the ground current is 2 amperes, the small ground current probably represents normal leakage. But if the neutral measures 40 amperes and the green ground measures 20 amperes, this probably indicates that there are improper neutral-ground connections. The smaller the ratio of neutral-to-ground current, the more likely it is that neutral-ground binds exist.

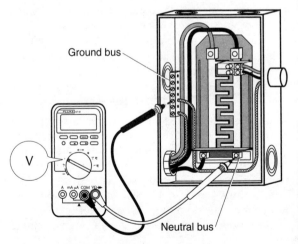

12-5 Measuring neutral current.

All neutral-to-ground connections not permitted by the National Electrical Code should be removed. This will improve both safety and power quality.

Overheated Terminals and Connections

Poor connections and loose terminations increase circuit impedance and thus voltage drop. They can also cause hard-to-diagnose intermittent problems, such as circuits that cycle ON and OFF unpredictably (a loose connection may open when it heats up, and then close again when it cools down).

"Hot spots" indicating possible poor connections and terminations can often be found using hand-held infrared probes or wands. Visual inspection may also be useful. A preventive maintenance program of checking and tightening conductor connections on a regular basis can help prevent this type of problem before it occurs.

Circuit Breakers

Although molded-case circuit breakers typically have long service lives, contacts and springs can wear out, particularly when the device has tripped frequently or been used as a switch to turn equipment or circuits ON and OFF. As with other poor connections, marginal breakers increase circuit impedance and voltage drop. Overheating due to their poor internal connections may also lead to "nuisance tripping."

Measure voltage drop across the circuit breaker, from line side to load side, to determine the condition of internal components (see Figure 12-6). If voltage drop exceeds 100 millivolts (mV), the branch circuit breaker should be replaced. Readings in the 35 to 100-mV range should be noted and those breakers rechecked at regular intervals.

Figure 12-7 summarizes power quality troubleshooting recommendations.

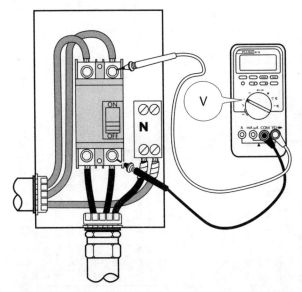

12-6 Measuring voltage drop across circuit breaker.

Symptoms	Probable Cause	Action or Items to Check
Low voltage levels at panelboard or service entrance equipment.	Utility supply voltage too low.	Consult utility.
	Transformer tap settings too low.	Use higher voltage taps.
	Loose connection in feeder service conductors.	Check connections.
Voltage sag coincides with current surge, when measured at panelboard.	Downstream load with high inrush current, such as motor(s) or incandescent lighting.	Consider feeding sensitive loads from other circuits or panelboards.
Voltage surge coincides with current decrease, when measured at panelboard.	Upstream source-related disturbance.	Consult utility.

12-7 Troubleshooting chart for power quality problems.

Symptoms	Probable Cause	Action or Items to Check
Significant neutral current on three-phase feeder.	Unbalanced loading on different phases of panelboard.	Balance panelboard.
Current on neutral of three-phase feeder equals or exceeds phase currents.	Harmonics generated by non-linear loads.	Increase size of feeder neutral conductor.
Current on shared neutral of multiwire branch of circuit equals or exceeds phase currents.	Harmonics generated by non-linear loads.	Use individual two-wire branch circuits.
Neutral-to-ground potential exceeds 0.5 volt.	Neutral-to-ground connections at panels other than service entrance equipment.	Remove improper neutral-to-ground connections.

12-7 Troubleshooting chart for power quality problems. *(Continued)*

Symptoms	Probable Cause	Action or Items to Check
Low voltage levels at receptacles or utilizations equipment.	Long branch circuit runs.	Install oversized conductors to compensate for voltage drops.
	Loose connections in branch circuits.	Check and tighten connections.
Voltage drop across circuit breaker exceeds 100 millivolts, from load to line side.	Worn circuit breaker contacts and springs.	Replace circuit breaker.

12-7 Troubleshooting chart for power quality problems. *(Continued)*

APPENDIX
Electrical Essentials

Electricity is basically the flow of electrons—tiny atomic particles. These particles are found in all atoms. Atoms of some metals such as copper and aluminum have electrons that are easily pushed and guided into a stream. When a coil of metal wire is turned near a magnet, or vice versa, electricity will flow in the wire. This principle is made use of in generating plants; water or steam is used to turn turbines which rotate electromagnets that are surrounded by huge coils of wire. The push transmitted to the electrons by the turbine/magnet setup is measured in units called *volts*. The quantity of the flow of electricity is called *current* and it is measured in *amperes* or *amps*.

Multiply volts by amps and you get *volt-amperes* or *watts*—the power or amount of work that electricity can do. Electrical appliances and motors have certain wattage requirements depending on the task they are expected to perform. For convenience, we can use *kilowatts* (1 kW equals 1000 W) when speaking of power production or power needs. A power plant produces kilowatts which are sold to users by the *kilowatt-hour*. For example, a 100-W lamp left on for 10 hours uses 1 kWh of electricity.

Resistance, or the opposition to the flow of electricity, is another term that will be covered in this appendix. In general, a conductor enhances the flow of electricity while resistance impedes or stops the flow of electricity. Therefore, insulators are constructed of materials offering a high resistance to the flow of electricity.

Measuring Electricity

The three basic terms used to measure electricity are:

- Electromotive force—measured in volts
- Current—measured in amperes
- Resistance—measured in ohms

Electromotive force or **voltage** is the force that causes electrons to flow. The unit of measurement of voltage is volts. In equations, voltage is represented by the letter *E*, which stands for electromotive force. Voltage is measured with a *voltmeter*.

Current is the rate at which electrons flow in a circuit. A current of one ampere is said to flow when one coulomb of charge passes a point in 1 second. One coulomb is equal to the charge of $6.28 / 10^{18}$ electrons. Current is measured in units called amperes or amps. In equations, current is represented by the letter *I* which stands for *intensity* of current.

In many cases, the ampere is too large a unit for measuring current. Therefore, the *milliampere* (mA) or the *microampere* (µA) is used. A milliampere equals one-thousandth of an ampere, while the microampere equals one-millionth of an ampere. The device used to measure current is called an *ammeter*.

Resistance is the opposition to the flow of current in a circuit. All circuits have some resistance and the amount of resistance is measured in ohms. In equations, resistance is represented by the letter R or the Greek letter omega (Ω). A conductor has one ohm of resistance when an applied potential of one volt produces a current of one ampere.

Opposition to current flow (resistance) is greater in a material with fewer free electrons. Consequently, the resistance of a material is determined by the number of free electrons available in a material. It is also determined by cross-sectional area. If the cross-sectional area of a conductor is increased, a greater quantity of electrons are available for movement through the conductor. Therefore, a larger current will flow for a given amount of applied voltage. If the cross-sectional area of a conductor is decreased, the number of available electrons decreases and, for a given applied voltage, the current through the conductor decreases.

Length is also a factor which determines the resistance of a conductor. The longer a conductor is, the more energy is lost to heat. Therefore, if the length of a conductor is increased, the resistance of that conductor increases.

Direct Current

The common flashlight is an example of a basic electric circuit. It contains a source of electrical energy (the dry cells in the flashlight), a load (the bulb) which changes the electrical energy into light energy, and a switch to control the energy delivered to the load.

The technician's main aid in studying circuits is the schematic diagram. A schematic diagram is a "picture" of the circuit that uses symbols to represent the various circuit components and lines to connect these components. Basic symbols used in schematic diagrams are shown in Figure A-1. Refer to this symbol list frequently as you review the various schematic diagrams in the remaining pages of this appendix.

Figure A-2 shows a pictorial representation of a common flashlight (A), along with the appropriate symbols, at the approximate locations, for the bulb, switch, and battery.

The schematic in (B) shows the flashlight in the OFF or deenergized state. The switch (S1) is open. There is no complete path for current (I) through the circuit, and the bulb does not light. In (C) however, switch S1 is closed (moved to the ON position) and current flows in the direction of the arrows from the negative terminal of the battery, through the switch (S1), through the lamp and back to the positive terminal of the battery. With the switch closed, the path for current is complete and current will continue to flow until the switch (S1) is moved to the open (OFF) position, or the battery is completely discharged.

Ohm's Law

Ohm proved by experimentation the relationship between current, voltage, and resistance (called Ohm's law) is stated as follows:

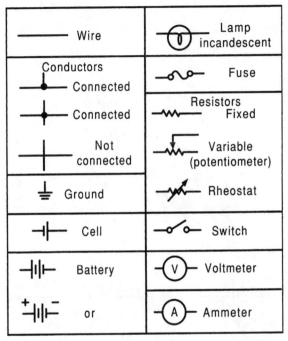

A-1 Basic symbols used in schematic diagrams.

The current in a circuit is directly proportional to the applied voltage and inversely proportional to the circuit resistance.

Ohm's law may be expressed as an equation:
$$I = \frac{E}{R}$$

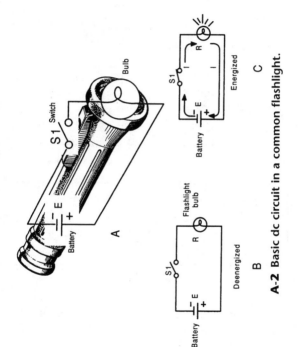

A-2 Basic dc circuit in a common flashlight.

where: I = current in amperes
E = voltage in volts
R = resistance in ohms

In using Ohm's law, if any two of the variables are known, the unknown can be found. For example, if current (I) and voltage (E) are known, resistance (R) can be determined as follows:

Step 1. Use the basic equation:

$$I = \frac{E}{R}$$

Step 2. Remove the divisor by multiplying both sides by R:

$$R \times I = \frac{E}{R} \times \frac{R}{I}$$

Step 3. Note result of Step 2: $R \times I = E$

Step 4. To get R alone, divide both sides by I:

$$\frac{RI}{I} = \frac{E}{I}$$

Step 5. The basic equation, transposed for R, is:

$$R = \frac{E}{I}$$

Now let's put this equation to practical use. Refer to Figure A-3 and note that the voltage (E) is 10 V, and the current (I) equals 1 A. Solve for R, using the equation just explained.

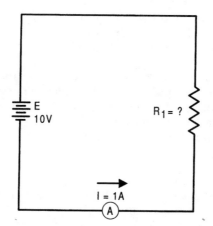

A-3 DC circuit with unknown resistance.

Step 1. Insert the known values in the equation:

$$R = \frac{10\text{ V}}{1\text{A}} = 10\ \Omega$$

The basic Ohm's law equation can also be solved for voltage (*E*).

Step 2. Use the basic equation:

$$I = \frac{E}{R}$$

Step 3. Multiply both sides by *R*:

$$I \times R = \frac{E}{R} \times \frac{R}{1}$$

Step 4. Note the results of Step 2.

$$E = I \times R$$

Now let's use this equation to find the voltage in the circuit shown in Figure A-4, where the amperage equals 0.5 A, and the resistance equals 45 Ω.

$$E = I \times R$$
$$E = 0.5 \text{ A} \times 45 \text{ Ω}$$
$$E = 22.5 \text{ V}$$

The Ohm's law equation and its various forms may be readily obtained with the aid of the circle in Figure A-5. To determine the unknown quantity, first cover that quantity with a finger. The position of the

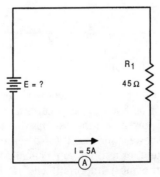

A-4 DC circuit with unknown voltage.

uncovered letters in the circle will indicate the mathematical operation to be performed. For example, to find current (*I*), cover the letter *I* with a finger. The uncovered letters indicate that *E* is to be divided by *R*, or $I = E/R$. To find the equation for *E*, cover *E* with your finger. The result indicates that *I* is to be multiplied by *R*, or $E = IR$. To find the equation for *R*, cover *R*. The result indicates that *E* is to be divided by *I*, or $R = E/I$.

Power

Power expresses the rate at which work is being done and is measured in watts or volt-amperes. Power in watts is equal to the voltage across a circuit multiplied by current through the circuit. This represents the rate at any given instant at which work is being done. The symbol *P* indicates electrical power, and the basic power equation is:

$$P = E \times I$$

where E = voltage and I = current.

The amount of power changes when either voltage or current, or both voltage and current, are caused to change. The power equation also has variations similar to those discussed previously for determining either voltage, current, or resistance in a circuit.

Four of the most important electrical quantities have been discussed thus far:

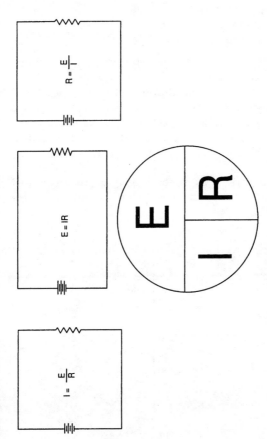

A-5 Ohm's law in diagram form.

- Voltage (*E*)
- Current (*I*)
- Resistance (*R*)
- Power (*P*)

Anyone involved in the electrical industry in any capacity must understand the relationships that exist among these quantities because they are used throughout this book and will be used throughout the technician's career. Figure A-6 is a summary of 12 basic equations that you should know. The four quantities *E*, *I*, *R*, and *P* are at the center of the circle. Adjacent to each quantity are three segments. Note that in each segment, the basic quantity is expressed in terms of two other basic quantities, and no two segments are alike.

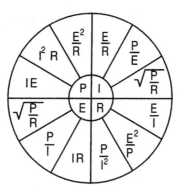

A-6 Summary of basic electrical equations.

Series DC Circuits

A series circuit is defined as a circuit that contains only one path for current flow. To compare the basic flashlight circuit (discussed previously) with a more complex series circuit, refer to Figure A-7. Remember that our basic flashlight circuit had only one lamp, while the more complex series circuit in Figure A-7 contains three lamps connected in series. The current in this circuit must flow through each lamp to complete the electrical path in the circuit. Each additional lamp offers added resistance. Consequently, in a series circuit, the total circuit resistance (R_T) is equal to the sum of the individual resistances.

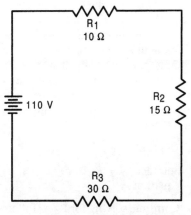

A-7 Solving for total resistance in a series circuit.

$$R_T = R_1 + R_2 + R_3 + \ldots + R_n$$

The series circuit in Figure A-7 consists of three resistors: one rated at 10 Ω, one at 15 Ω, and one at 30 Ω. A voltage source provides 120 V. What is the total resistance?

Step 1. Use the basic equation for finding resistance in a series circuit.

Step 2. Substitute known values in the equation.

$$R_T = 10 \text{ Ω} + 15 \text{ Ω} + 30 \text{ Ω}$$
$$R_T = 55 \text{ Ω}$$

In some applications, the total resistance is known and the value of one of the circuit resistors has to be determined. The former equation for finding resistance in a series circuit can be transposed to solve for the value of the unknown resistance.

For example, the series circuit in Figure A-8 has a total resistance of 40 Ω. Two of the resistors are rated at 10 Ω each, while the rating of resistor R_3 is unknown. Here's how to find the value of the unknown resistor.

Step 1. Use the basic equation.

Step 2. Subtract $R_1 + R_2$ from both sides of the equation.

Step 3. Continue solving for R_3 as follows:

$$R_T - R_1 - R_2 = R_3$$

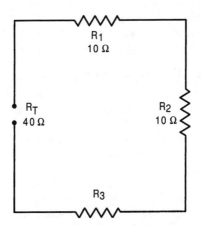

A-8 Calculating the value of one unknown resistance in a series circuit.

$$R_3 = R_T - R_1 - R_2$$
$$R_3 = 40\ \Omega - 10\ \Omega - 10\ \Omega$$
$$R_3 = 40\ \Omega - 20\ \Omega$$
$$R_3 = 20\ \Omega$$

Current in a Series Circuit

Since there is only one path for current in a series circuit, the same current must flow through each component of the circuit. Ohm's law may be used to calculate the current in a series circuit if the voltage and resistance quantities are known.

The current flow through each component of a series circuit can be verified by inserting meters into the circuit at various points, as shown in Figure A-9. Upon examining these meters, each meter would be found to indicate the same value of current.

Voltage in a Series Circuit

The voltage drop across the resistor in a circuit consisting of a single resistor and a voltage source is the total voltage across the circuit and is equal to the applied voltage. The total voltage across a series circuit that consists of more than one resistor is also equal to the applied voltage, but consists of the sum of the individual resistor voltage drops.

In any series circuit, the sum of the resistor voltage drops must equal the source voltage. In the circuit shown in Figure A-10, a source voltage (E_T) of 20 V is

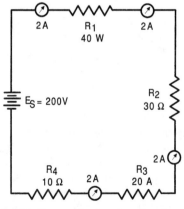

A-9 Current in a series circuit.

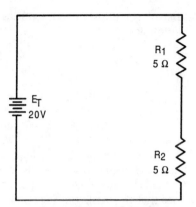

A-10 Calculating individual voltage drop in a series circuit.

dropped across a series circuit consisting of two 5-Ω resistors. The total resistance of the circuit (R_T) is equal to the sum of the two individual resistances, or 10 Ω. Using Ohm's law, the circuit current may be calculated as follows:

$$I_T = \frac{E_T}{R_T}$$

$$I_T = \frac{20 \text{ V}}{10 \text{ Ω}}$$

$$I_T = 2 \text{ A}$$

Since the value of the resistors is known to be 5 Ω each, and the current through the resistors is known to be 2 A, the voltage drops across the resistors can be calculated as follows:

$$E_1 = I_1 \times R_1$$
$$E_1 = 2 \text{ A} \times 5 \text{ }\Omega$$
$$E_1 = 10 \text{ V}$$

Since R_2 is the same ohmic value as R_1, and carries the same current, the voltage drop across R_2 also equals 10 V. Then, adding the voltage drops for R_1 and R_2, we obtain (10 + 10 =) 20 V, which is equal to the applied voltage in the circuit. From the previous explanation, we see that the total voltage in a dc series circuit may be obtained by using the following equation:

$$E_T = E_1 + E_2 + E_3 + \ldots + E_n$$

Power in a Series Circuit

Each of the resistors in a series circuit consumes power which is dissipated in the form of heat. In a series circuit, the total power is equal to the sum of the power dissipated by the individual resistors. The equation to find the total power in a series circuit follows:

$$P_T = P_1 + P_2 + P_3 + \ldots + P_n$$

Let's determine the total power in watts for the series circuit In Figure A-11. Note that this circuit has an applied potential of 120 V and three resistors are connected in series, each rated at 5 Ω, 10 Ω, and 15 Ω, respectively.

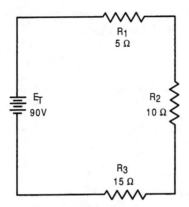

A-11 Solving for total power in a series circuit.

Step 1. Find the total resistance in the circuit.

$$R_T = R_1 + R_2 + R_3$$
$$R_T = 5\,\Omega + 10\,\Omega + 15\,\Omega$$
$$R_T = 30\,\Omega$$

Step 2. Determine the circuit current.

$$I = \frac{E_T}{R_T}$$

$$I = \frac{120\text{ V}}{30\,\Omega}$$

$$I = 4\text{ A}$$

Step 3. Use the power equation to calculate the power for each resistor.

$$P_{1(2)(3)} = I^2 \times R_{1(2)(3)}$$

Step 4. Calculate the power for resistor R_1.

$$P_1 = (4 \text{ A})^2 \times 5 \text{ }\Omega$$
$$P_1 = 80 \text{ W}$$

Step 5. Calculate the power for resistors R_2 and R_3 using the same steps as given in Step 4.

$$P_2 = 160 \text{ W}$$
$$P_3 = 240 \text{ W}$$

Step 6. Obtain the total power by adding all power values.

$$P_T = 80 \text{ W} + 160 \text{ W} + 240 \text{ W}$$
$$P_T = 480 \text{ W}$$

When the total source voltage and the total source amperage are known, the total wattage may be found by multiplying the volts times the amps (volt-amperes). Let's check the previous example with this method.

$$P_{\text{Source}} = E_{\text{Source}} = I_{\text{Source}}$$
$$P_{\text{Source}} = 120 \text{ V} \times 4 \text{ A}$$
$$P_{\text{Source}} = 480 \text{ VA (W)}$$

Kirchhoff's Voltage Law

In 1847, G. R. Kirchhoff extended the use of Ohm's law by developing a simple concept concerning the voltages contained in a series circuit loop. Kirchhoff's voltage law states:

> The algebraic sum of the voltage drops in any closed path in a circuit equals the supply voltage.

Kirchhoff's voltage law can be written as an equation as follows:

$$E_a + E_b + E_c + \ldots + E_n = 0$$

where E_a, E_b, etc., are the voltage drops around any closed circuit loop.

To set up the equation for an actual circuit, the following procedure is used:

Step 1. Assume a direction of current through the circuit.

Step 2. Using the assumed direction of current, assign polarities to all resistors through which the current flows.

Step 3. Place the correct polarities on any sources included in the circuit.

Step 4. Starting at any point in the circuit, trace around the circuit, writing down the amount and polarity of the voltage across each component in succession. The polarity used is the sign after the assumed current has passed through the component. Stop when the point at which the trace was started is reached.

Step 5. Place these voltages, with their polarities, into the equation and solve for the desired quantity.

To place the above procedures in use, assume that three resistors are connected in series with a 50-V source. What is the voltage across the third resistor if the voltage drops across the first two resistors are 25 V and 15 V, respectively?

Step 1. Draw a diagram such as the one shown in Figure A-12.

Step 2. Draw an arrow indicating the assumed direction of current flow.

Step 3. Using the current direction arrow as made in Step 2, mark the polarity (− or +) at each end of each resistor and also on the terminals of the source.

Step 4. Starting at point A, trace around the circuit in the direction of current flow, recording the voltage and polarity of each component.

Step 5. Starting at point A and using the components from the circuit, we have:

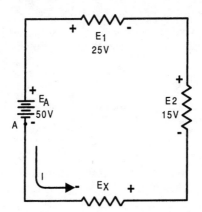

A-12 Determining unknown voltage in a series circuit.

$$(+Ex) + (+E_2) + (+E_1) + (-E_a) = 0$$

Step 6. Substitute known values in the equation from the circuit.

$$Ex + 15\text{ V} + 25\text{ V} - 50\text{ V} = 0$$
$$Ex - 10\text{ V} = 0$$
$$Ex = 10\text{ V}$$

The unknown voltage (Ex) is found to be 10 V.

Solving for Unknown Current

Using the same procedure as above, problems may be solved in which the current is the unknown quantity.

For example, let's assume that a series circuit has a source voltage of 60 V and contains three resistors of 5 Ω, 10 Ω, and 15 Ω. Find the circuit current.

Step 1. Draw and label the circuit as shown in Figure A-13.

Step 2. Start at any point and write out the loop equation.

$$E_2 + E_1 + E_A + E_3 = 0$$

Step 3. Since $E = IR$, substitute known values in the equation.

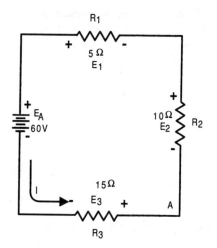

A-13 Correct direction of assumed current.

$$(I \times R_2) + (I \times R_1) + E_A + (I \times R_3) = 0$$

$$(I \times 10\ \Omega) + (I \times 5\ \Omega) + (-60\ \text{V}) + (I \times 15\ \Omega) = 0$$

Step 4. Combine like terms.

$$(I \times 30\ \Omega) + (-60\ \text{V}) = 0$$

$$I = \frac{60\ \text{V}}{30\ \Omega}$$

$$I = 2\ \text{A}$$

Since the current obtained in the preceding calculation is a positive 2 A, the assumed direction of current was correct. However, if the calculation had been a negative value, the assumed direction of current flow would be incorrect. Even if the wrong current direction is assumed, the amount of current in the calculation will be the same. The polarity, however, is negative if the wrong current direction is chosen. In this case, all that is required is to reverse the direction of the assumed current flow. However, should it be necessary to use this negative current value in further calculations on the circuit using Kirchhoff's law, the negative polarity should be retained in the calculations.

Parallel DC Circuits

A *parallel circuit* is defined as one having more than one current path connected to a common voltage source. Parallel circuits, therefore, must contain two or more resistances which are not connected in series. An example of a basic parallel circuit is shown in Figure A-14.

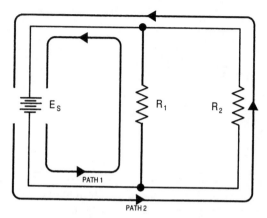

A-14 A basic parallel circuit.

Start at the voltage source (E_S) and trace counterclockwise around the circuit. Two complete and separate paths can be identified in which current can flow. One path is traced from the source, through resistance R_1, and back to the source. The other path is from the source, through resistance R_2, and back to the source.

Voltage in a Parallel Circuit

The source voltage in a series circuit divides proportionately across each resistor in the circuit. However, in a parallel circuit, the same voltage is present in each branch. In Figure A-14 this voltage is equal to the applied voltage (E_S) and can be expressed in the following equation:

$$E_S = E_{R_1} = E_{R_2}$$

Voltage measurements taken across the resistors of a parallel circuit are illustrated in Figure A-15. Each voltmeter indicates the same amount of voltage. Also note that the voltage across each resistor in the circuit is the same as the applied voltage.

For example, assume that the current through a resistor of a parallel circuit is known to be 4.5 mA and the value of the resistor is 30,000 Ω (30 kΩ). What is the source voltage?

The circuit in question is shown in Figure A-16 and the source voltage may be found by using the basic Ohm's law equation:

$$E = IR$$

Substituting the known values in the equation, we have the following:

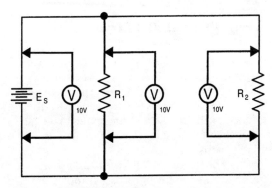

A-15 Voltage comparison in a parallel circuit.

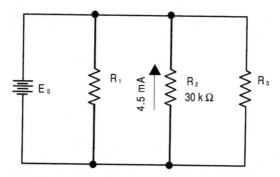

A-16 Finding source voltage in a parallel circuit.

$$E_{R_2} = 0.0045 \text{ A} \times 30{,}000 \text{ }\Omega$$
$$E_{R_2} = 135 \text{ V}$$

Since the source voltage is equal to the voltage of a branch, 135 V is the source voltage and is also the voltage applied to each branch of this circuit.

Current in a Parallel Circuit

Ohm's Law states that the current in a circuit is inversely proportional to the circuit resistance. This fact is true in both series and parallel circuits.

There is a single path for current in a series circuit. The amount of current is determined by the total resistance of the circuit and the applied voltage. In a parallel circuit, the source current divides among the available paths.

Part (A) of Figure A-17 shows a basic series circuit. Here, the total current must pass through the single resistor (R_1). Note that the applied voltage equals 50 V and the resistance of R_1 is 10 Ω. The amount of current can be determined by using Ohm's law and is calculated as follows:

$$I = \frac{E}{R}$$

$$I_T = \frac{50 \text{ V}}{10 \text{ Ω}}$$

$$I_T = 5 \text{ A}$$

Part (B) of Figure A-17 shows the same resistor (R_1) with a second resistor (R_2) of equal value connected in parallel across the voltage source. When Ohm's law is applied, the current flow through each resistor is found to be the same as the current through the single resistor in part (A).

$$I = \frac{E}{R}$$

$$E_S = E_{R_1} = E_{R_2}$$

$$I_{R_1} = \frac{50 \text{ V}}{10 \text{ Ω}}$$

$$I_{R_1} = 5 \text{ A}$$

$$I_{R_2} = \frac{50 \text{ V}}{10 \text{ Ω}}$$

$$I_{R_2} = 5 \text{ A}$$

It is apparent that if there is 5A of current through each of the two resistors, there must be a total current of 10 A drawn from the source.

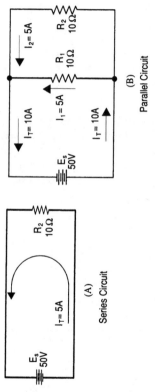

A-17 Analysis of current in a series and parallel circuit.

The total current of 10 A leaves the negative terminal of the battery and flows to the first junction point, where it divides into two currents of 5 A each. These two currents flow through their respective resistors and rejoin at the second junction. The total current then flows back to the positive terminal of the source. The source supplies a total current of 10 A and each of the two equal resistors carries one-half of the total current.

Each individual current path in the circuit in Figure A-17(B) is referred to as a *branch*. Each branch carries a current that is a portion of the total current. Two or more branches form a *network*.

From the previous explanation, the characteristics of current in a parallel circuit can be expressed in terms of the following general equation:

$$I_T = I_1 + I_2 + \ldots + I_n$$

Kirchhoff's Current Law

The division of current in a parallel network follows a definite pattern. This pattern is described by Kirchhoff's current law which states:

> The algebraic sum of the currents entering and leaving any junction of conductors is equal to zero.

This law can be stated mathematically as:

$$I_a + I_b + \ldots + I_n = 0$$

where: I_a, I_b, etc. are the currents entering and leaving the junction. Currents entering the junction are considered to be positive and currents leaving the junction are considered to be negative. When solving a problem using Kirchhoff's current law, the currents must be placed into the equation with the proper polarity signs attached.

Now let's use Kirchhoff's current law to solve for the value Of I_3 in Figure A-18.

The known values are first substituted in Kirchhoff's current law equation.

$$I_1 + I_2 + I_3 + I_4 = 0$$

$$10 \text{ A} + (-3 \text{ A}) + I_3 + (-5 \text{ A}) = 0$$

$$I_3 + 2 \text{ A} = 0$$

$$I_3 = -2 \text{ A}$$

Current I_3 has a value of 2 A, and the negative sign shows it to be a current leaving the junction.

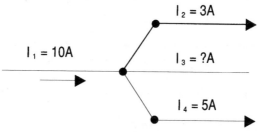

A-18 Circuit with four current values.

Resistance In Parallel Circuits

Figure A-19 shows two resistors connected in parallel across a 5-V battery. Each has a resistance value of 10 Ω. A complete circuit consisting of two parallel paths is formed and current flows as shown.

Computing the individual currents show 0.5 A flows through each resistor. The total current flowing from the battery to the junction of the resistors, and returning from the resistors to the battery, is equal to 1 A.

The total resistance of the circuit can be calculated by using the values of total voltage (E_T) and total current (I_T).

$$R = \frac{E}{I}$$

$$R_T = \frac{5 \text{ V}}{1 \text{ A}}$$

$$R_T = 5 \text{ Ω}$$

This computation shows the total resistance to be 5 Ω, one-half the value of either of the two resistors.

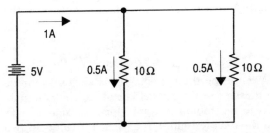

A-19 Two equal resistors connected in parallel.

> The total resistance of a parallel circuit is smaller than any of the individual resistors.

In other words, the total resistance in a parallel circuit is not the sum of the individual resistor values as was the case in a series circuit.

There are several methods used to determine the total or equivalent resistance of parallel circuits. Where all resistors in the circuit are of the same value the following simple equation may be used:

$$R_T = \frac{R}{N}$$

where: R_T = total parallel resistance
R = ohmic value of one resistor
N = number of resistors

The equation is valid for any number of parallel resistors of equal value.

The total resistance of parallel circuits can be found if the individual resistance values are known along with the source voltage. The following equation is the most common:

$$\frac{1}{R_T} = \frac{1}{R_1} + \frac{1}{R_2} + \frac{1}{R_3} + \cdots \frac{1}{R_n}$$

When using the preceding equation to determine the resistance in a parallel circuit, convert the fractions to a common denominator. For example, let's find the total resistance in a parallel circuit with two resistors rated at 3 Ω and 6 Ω, respectively.

$$\frac{1}{R_T} = \frac{1}{3\,\Omega} + \frac{1}{6\,\Omega}$$

$$\frac{1}{R_T} = \frac{2}{6\,\Omega} + \frac{1}{6\,\Omega}$$

$$\frac{1}{R_T} = \frac{3}{6\,\Omega}$$

$$\frac{1}{R_T} = \frac{1}{2\,\Omega}$$

Since both sides are reciprocals (divided into one) disregard the reciprocal function.

$$R_T = 2\,\Omega$$

When only two resistors, each of either the same or different values, are in a parallel circuit, the following equation may be used to find the total resistance in the circuit.

$$R_T = \frac{R_1 \times R_2}{R_1 \times R_2}$$

Using the above equation, what is the total resistance in the circuit shown in Figure A-20?

$$R_T = \frac{20\,\Omega \times 30\,\Omega}{20\,\Omega \times 30\,\Omega}$$

$$R_T = \frac{600}{50}\,\Omega$$

$$R_T = 12\,\Omega$$

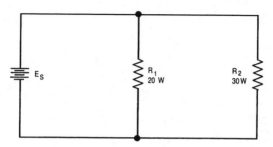

A-20 Parallel circuit with two unequal resistors.

Equivalent Circuits

In dealing with electrical circuits, it is sometimes necessary to reduce a complex circuit into a simpler form. Any complex circuit consisting of resistances can be redrawn (reduced) to a basic equivalent circuit containing the voltage source and a single resistor representing total resistance. This process is called reduction to an *equivalent circuit*.

Figure A-21 shows a parallel circuit with three resistors of equal value and the redrawn equivalent circuit. The parallel circuit shown in part (A) shows the original circuit. To create the equivalent circuit, first calculate the total resistance in the circuit.

$$R_T = \frac{R}{N}$$

$$R_T = \frac{45 \ \Omega}{3}$$

$$R_T = 15 \ \Omega$$

$$\frac{1}{R_T} = \frac{1}{3\,\Omega} + \frac{1}{6\,\Omega}$$

$$\frac{1}{R_T} = \frac{2}{6\,\Omega} + \frac{1}{6\,\Omega}$$

$$\frac{1}{R_T} = \frac{3}{6\,\Omega}$$

$$\frac{1}{R_T} = \frac{1}{2\,\Omega}$$

Since both sides are reciprocals (divided into one) disregard the reciprocal function.

$$R_T = 2\,\Omega$$

When only two resistors, each of either the same or different values, are in a parallel circuit, the following equation may be used to find the total resistance in the circuit.

$$R_T = \frac{R_1 \times R_2}{R_1 \times R_2}$$

Using the above equation, what is the total resistance in the circuit shown in Figure A-20?

$$R_T = \frac{20\,\Omega \times 30\,\Omega}{20\,\Omega \times 30\,\Omega}$$

$$R_T = \frac{600}{50}\,\Omega$$

$$R_T = 12\,\Omega$$

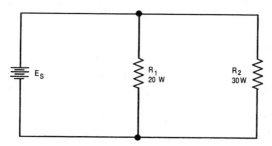

A-20 Parallel circuit with two unequal resistors.

Equivalent Circuits

In dealing with electrical circuits, it is sometimes necessary to reduce a complex circuit into a simpler form. Any complex circuit consisting of resistances can be redrawn (reduced) to a basic equivalent circuit containing the voltage source and a single resistor representing total resistance. This process is called reduction to an *equivalent circuit*.

Figure A-21 shows a parallel circuit with three resistors of equal value and the redrawn equivalent circuit. The parallel circuit shown in part (A) shows the original circuit. To create the equivalent circuit, first calculate the total resistance in the circuit.

$$R_T = \frac{R}{N}$$

$$R_T = \frac{45\ \Omega}{3}$$

$$R_T = 15\ \Omega$$

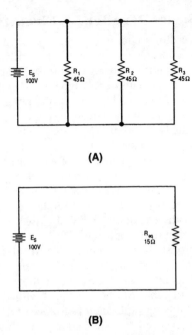

A-21 Parallel circuit (A) with equivalent circuit (B).

Once the equivalent resistance is known, a new circuit is drawn consisting of a single resistor (to represent the equivalent resistance) and the voltage source, as shown in part (B).

Series-Parallel DC Circuits

In the preceding sections, series and parallel dc circuits have been considered separately. Electricians often encounter circuits consisting of both series and parallel elements. A circuit of this type is referred to as a *combination circuit*. Solving for the quantities and elements in a combination circuit is simply a matter of applying the laws and rules discussed up to this point.

The basic technique used for solving dc combination circuit problems is the use of equivalent circuits. To simplify a complex circuit to a simple circuit containing only one load, equivalent circuits are substituted (on paper) for the complex circuit they represent—the technique briefly discussed in the preceding section of this appendix.

To demonstrate the method used to solve combination circuit problems, refer to the circuit in Figure A-22(A). Examination of this circuit shows that the only quantity that can be computed with the given information is the equivalent resistance of R_2 and R_3. Since only two resistors are contained in this part of the circuit, and since these resistors are connected in parallel, the product over the sum equation may be used to obtain the total resistance for this portion of the circuit.

$$R_T = \frac{R_2 \times R_3}{R_2 + R_3}$$

$$R_T = \frac{20 \; \Omega \times 30 \; \Omega}{20 \; \Omega \; + 30 \; \Omega}$$

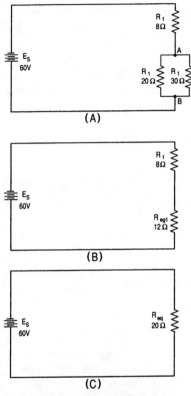

A-22 Steps in reducing a combination circuit to its simplest form.

$$R_T = \frac{600}{50} \, \Omega$$

$$R_T = 12 \, \Omega$$

Now that the equivalent resistance for R_2 and R_3 has been calculated, the circuit can be redrawn as a series circuit as shown in Figure A-22(B). The total resistance for the entire circuit may now be calculated as follows:

$$R_T = 8\Omega + 12 \, \Omega$$

$$R_T = 20 \, \Omega$$

The original circuit can be redrawn with a single resistor that represents the equivalent resistance of the entire circuit as shown in Figure A-22(C).

$$I_T = \frac{60 \text{ V}}{20 \, \Omega}$$

$$I_T = 3 \text{ A}$$

Summary

- A basic electric circuit consists of a source of electrical energy connected to a load. The load uses the energy and changes it to a useful form.
- A schematic diagram is a "picture" of a circuit that uses symbols to represent components. The space required to depict an electrical or electronic circuit is greatly reduced by the use of a schematic diagram.

- Ohm's law can be transposed to find one of the values in a circuit if the other two values are known.
- Equal current flows through each part of a series circuit.
- The total resistance of a series circuit is equal to the sum of the individual resistances.
- The total voltage in a series circuit is equal to the sum of the individual voltage drops (Kirchhoff's voltage low).
- The voltage drop across a resistor in a series circuit is proportional to the ohmic value of the resistor.
- The total power in a series circuit is equal to the sum of the individual power used by each circuit component.

Index

Note: boldface numbers indicate illustrations:

ac circuits, voltmeter testing of, 9, 10
ac motors
  megohmmeter testing of, 21–22, **23**
  ohmmeter testing of, 18–19, **19**
alternators, synchroscopes for, 29–30, **31**
ammeters, 1, 2–8, **2**
  applications for, 6–7
  current measurement using, 3, 6–7
  current multipliers for, 5, **5**
  extending range of, 4, **4**
  precautions when using, 5–6
  recording-type, 7–8, **8**
amperes, 241, 242
analog meters, 1–25
  ammeters, 1, 2–8, **2**
  digital meters vs. 1–2
  megohmmeters (meggers), 1, 17–25
  resistance testers (*See* megaohmmeters)
  voltmeters, 1, 8–16
angular-contact ball bearings, **195**, 196

ball-thrust bearings, **195**, 197
bearings, motor (*See* motor bearings)
branch circuits, 271
branch-circuit type panel board, **232**

cable-length meters, 34
capacitance, digital multimeters (DMM) measurement using, 44, **46**
centrifugal switches, defective, in split-phase motor, 128–129
circuit breakers
  megohmmeter testing, 22
  power quality problems and, 236–237, **237**
  voltage drop in, 236–237, **237**
combination circuits, 278–280, **279**
conductors, 242, 243
contactors (*See* relays and contactors)
continuity testing
  digital multimeter (DMM) for, 42, **45**, 54–55
  three-phase, delta-wound motors, 140
current, 241, 242, 244, 252
  current loading in, 231

283

current, (*Cont.*):
  Kirchhoff's current law and, 271–272
  measurement of, 3, 6–7, 42, **43**
  parallel circuits, 268–271, **270**
  series circuits, 255–256, **256**
  solving for unknown, 263–265, **264**
current loading, 231
current multipliers, ammeter, 5, **5**
cylindrical-roller bearings, **195**, 197

dc circuits
  parallel, 265–275
  series, 253–255, **253**, **255**
  voltmeter testing of, 9
dc motors, megohmmeter testing of, 19–21, **21**
de-energized lead reading in motors, 107
digital meters vs analog meters, 1–2
digital multimeters (DMM), 2, 37–49
  accessories for, 41
  accuracy of, 37–38
  capacitance measurement using, 44, **46**
  construction features of, 39–40
  continuity testing using, 42, **45**, 54–55
  current measurement using, 42, **43**
  diode testing with, 45, **47**
  displays for, 38–39
  energy-rating categories for, 48
  freeze or capture mode of, 39

digital multimeters (DMM), (*Cont.*):
  frequency measurement using, 44
  function selection in, 40
  hold button feature of, 39
  inputs for, 40–41
  recording test results with, 55
  resistance measurement using, 42, **44**
  safety features of, 46–49
  test leads for, 40–41
  voltage measurement using, 41–42, **43**
diodes, digital multimeters (DMM) to test, 45, **47**
direct current (dc), 243–244, **246**
double row bearings, 194
double-row, deep groove ball bearings, **195**, 196

electrical service, industrial, **13**
electromotive force, 242
electron flow and electricity, 241
equivalent circuits, 276–278, **277**

fluorescent fixtures, 65–82, **66**
  black spot near center, 77
  blackening of ends, 71, 72, 73, 74, 75
  blackening of tube, 75, 76, 77
  blinking, 68, 69
  brownish rings around tube, 78
  color differences, 82
  dark section in tube, 81
  dark streaks on tube, 78
  decreased light output, 81, 82
  ends remain lit, 69
  humming noise, 81
  no starting, 70

fluorescent fixtures, (*Cont.*):
  preheat lamps, 65
  radio interference, 80
  rapid start circuits, 65
  rapid start circuits in, 67
  short life, 73, 74
  slow starting, 70, 71, 72
  swirling or fluttering light, 79, 80
  trigger start circuits, 65
  troubleshooting charts for, 68–82
  unequal brilliancy, 82
footcandle meter, 32–33
frequency, digital multimeters (DMM) measurement using, 44
frequency meters, 27–28
fuses, voltmeter testing of, 16, **16**

generators, megohmmeter testing of, 19–21, **21**
ground faults
  transformers, dry-type, 60–61
  voltmeter testing using, 12–16
ground resistance, megohmmeter testing, 23–25, **24**
ground reversal in motors, 129
ground testing
  motors, 105
  power quality problems and, 234–236, **235**
grounded coils in motors, 107–108, 111
growler to test motor coils, 112–114, **113**

harmonics, 231, 233
high-intensity discharge (HID) lamps, 89–101, **90**

high-intensity discharge (HID) lamps, (*Cont.*):
  arc tube swollen, broken, cracked, 99–100
  blackened inner arc tube, 101
  blinks out, 94
  broken or cracked outer bulb, 98, 99
  dim light, 91, 92, 93
  flickering light, 95, 96
  green or brown bulbs, 95
  lamp won't screw in, 97–98
  loose base, 101
  melted wires in arc tube, 100
  no light, 91, 92
  noisy ballast, 96–97
  overcurrent, 98
  radio interference, 96
  rattle, 101
  shape of bulbs in, 90
  split ends of arc tube, 100
  strobing light, 95, 96
  sunburn or suntan to people working near, 97
  troubleshooting charts for, 91–101

incandescent lamps, 83–88, **84**
  broken lamp, 87
  dim light, 86
  no light, 86
  quartz lamps, 83–85, **84**
  short life, 86, 87
  troubleshooting charts for, 86–88
infinity reading in tests of motors, 107
infrared sensing device, 34
insulation failure
  motors, 104–105
  transformers, dry-type, 63, 64

insulators, 242
intermittent faults, 54

kilowatt-hours, 241
kilowatts, 241
Kirchhoff, G.R., 261
Kirchhoff's current law, 271–272
Kirchhoff's voltage law, 261–265

low-voltage test, 11–12
  megaohmmeter, 18–19, **19**, **20**
lubrication of bearings, 198–202

main distribution panel board, **228**
megaohmmeters (meggers), 1, 17–25, **18**
  ac motor testing, 18–19, **19**, 21–22, **23**
  circuit breaker testing, 22
  dc motor testing, 19–21, **21**
  generator testing, 19–21, **21**
  ground resistance testing, 23–25, **24**
  low-voltage testing using, 18–19, **19**, **20**
  safety switch testing, 22–23
  switch testing, 22–23
mercury vapor lamps (*See* high-intensity discharge (HID) lamps)
microamperes, 242
milliamperes, 242
motor bearings, 193–209
  air gaps in, 203
  angular-contact ball, **195**, 196
  assembly/disassembly of, 198–202

motor bearings, (*Cont.*):
  ball-thrust, **195**, 197
  cylindrical-roller, **195**, 197
  dirt in, 202
  double row, 194
  double-row, deep groove ball, **195**, 196
  double-shielded, 193
  end play in, 203–204
  lubrication of, 198–202
  maintenance of, 198–202
  oil levels for, 204
  oil-ring sleeve, 194
  open, 193, 194
  overgreasing problems in, 202
  overheating in, 205–209
  plain sleeve, 194
  relief hole in, 202
  sealed, 194
  self-aligning ball, 195–196, **195**
  shaft tests, 204
  single row, deep-groove ball, **195**, 196
  single-shielded, 193, 194
  sleeve type, 194, 202
  spherical-roller thrust, **195**, 197–198
  spherical-roller, **195**, 197
  tapered-roller, **195**, 198
  testing, 203–204
  troubleshooting charts for, 205–209
  types of, in motors, 195
motors, 103–192
  bearings in (*See* motor bearings)
  bent shaft, bearings out of line, 127
  brush wear excessive, 175–178
  circuit identification in, 133–138, **134**, **135**, **139**

motors, (*Cont.*):
    continuity testing in, 140
    de-energized lead reading in, 107
    defective centrifugal switches, 128–129
    dirt and dust in, 104, 150–151
    eyebolt lifting capacity of, 132
    fast speed, 158–160
    gains speed and won't slow, 160–161
    ground reversal, 129
    ground testing in, 105
    grounded coils in, 107–108, 111
    growlers to test coils in, 112–114, **113**
    hot armature, 165–166
    hot ball bearings in, 147–149
    hot bearings in, 145–147
    hot commutator, 166–168
    hot fields, 168–169
    identification of, 133–138, **134**, **135**, **139**
    infinity reading in tests of, 107
    insulation failure in, 104–105
    insulation-resistance tests for, 106–107
    lifting and moving of, 130–131
    magnetic noise, 192
    moisture problems in, 104, 111, 130, 132, 151–152
    NEMA Standard for identification of, 133–138, **134**, **135**, **139**
    no commutation, 171–175
    no start, 153–154, 182
    noisy operation, 178–180, 192
    oil leaks in, 149–150
    open circuits in, 105, 128

motors, (*Cont.*):
    open coils in, 114
    overheating, 119–120, 163–164, 186–188
    resistance tests for, 106–107
    reversed connections in, 114–116, 129
    reversed phase in, 116–117
    rotor in, 103
    safety switch, 400A, **15**
    scraping noise, 192
    short circuits in, 105–106, 129–130
    shorted coils in, 112–114, **113**
    slow acceleration, 185–186
    slow speed, 122–123, 155–157, 161–162, 183–185
    sparking at brushes, 171–175
    split-phase motors, 117–130
        bent shaft, bearings out of line, 127
        defective centrifugal switches, 128–129
        ground reversal, 129
        magnetic noise, 126
        no start, 121–122
        open circuits, 128
        overheating, 119–120
        reversed connections, 129
        short circuits, 129–130
        slow acceleration, 118
        slow speed, 118, 122–123
        stalling motor, 120–121
        starts, then quits, 122
        tight bearings, 127
        unbalanced line current, 125–126
        vibration, 124–125
        worn bearings, 127
        wrong rotation, 118

motors, (*Cont.*):
  stalls, 120–121, 181–182
  starts, then quits, 183
  starts, then reverses direction, 154–155
  stator problems in, 103–104
  storage of, 130–132
  three-phase, delta-wound motors, 140–143, **141**
    continuity testing in, 140
    voltage testing in, 141–143, **142**
  tight bearings, 127
  tools for troubleshooting, 106–107, 108–110
  transistorized stethoscope for testing, 106
  troubleshooting charts for, 118–130, 144–192
  unbalanced line current, 191
  unbalanced, 169–171
  vibration after corrections, 188–190
  vibration, 124–125, 169–171
  voltage testing in, 141–143, **142**
  windings in, 103, 104, 105
  worn bearings, 127
  wrong rotation, 186
multivapor lamps (*See* high-intensity discharge (HID) lamps)

NEMA Standard for identification of motors, 133–138, **134**, **135**, **139**
network of circuits, 271

Ohm's law, 243, 247–250, **251**, 261
oil-ring sleeve bearings, 194

open bearings, 194
open circuits
  motors, 105, 114, 128
  split-phase motors, 128
  transformers, dry-type, 57, 59–60, **59**
open coils in motors, 114

panel board, **228**
  branch-circuit type, **232**
  voltage measurement at, 229, **229**
parallel circuits, 265–275, **266**
  branch circuits in, 271
  current in, 268–271, **270**
  dc, 265–275
  Kirchhoff's current law and, 271–272
  network of circuits in, 271
  resistance in, 273–275, **273**, **276**
  series-parallel circuit, 278–280, **279**
  voltage in, 266–268, **267**, **268**
phase, reversed phase in motors, 116–117
phase-sequence indicators, 33–34
plain sleeve bearings, 194
power, 250, 252
  series circuits, 258–260, **259**
power quality analyzers, 34–36, **35**, 34
power quality problems, 227–240
  circuit breakers and, 236–237, **237**
  current loading in, 231
  drop in, 236–237, **237**
  grounding in, 234–236, **235**
  harmonics in, 231, 233
  multiwire branch circuits and harmonics in, 233–234

power quality problems, (*Cont.*):
  overheating in, 236
  sags in voltage, 230–231
  stability of voltage levels, 227, 229–231
  triplen harmonics in, 231, 233
  troubleshooting charts for, 238–240
  voltage levels in, 227, 229–231
  voltage measurement at panel board in, 229, **229**
power-factor meter, 28–29, **29**
preheat lamps in motors, 65

quartz lamps (*See also* incandescent lamps), 83–85, **84**

radio interference
  fluorescent lights, 80
  high-intensity discharge (HID) lamps, 96
rapid start circuits, motors, 65, 67
recording ammeter, 7–8, **8**
relays and contactors, 211–225, **212**, **213**
  chatter or hum, 223–225
  discolored contacts, 222–223
  failure to drop out, 218–221
  failure to pull in, 214–217
  failure to start with contactor closed, 217
  hot coil temperature, 225
  pitted contacts, 222–223
  time delay relays operate too fast, 221
  troubleshooting charts for, 214–225
resistance/resistors, 241, 242–243, 252

resistance/resistors, (*Cont.*):
  digital multimeters (DMM) measurement using, 42, **44**
  parallel circuits, 273–275, **273**, **276**
resistance testers, 1
  motors, 106–107
reversed connections in motors, 129
reversed phase in motors, 116–117
rotor in motors, 103

safety switch, megaohmmeter testing, 22–23
safety switch, 400A, **15**
sags in voltage, 230–231
schematics, symbols used in, **244**
sealed bearings, 194
self-aligning ball bearings, 195–196, **195**
series circuits
  current in, 255–256, **256**
  dc, 253–255, **253**, **255**
  power in, 258–260, **259**
  voltage in, 256–258, **257**
series-parallel circuit, 278–280, **279**
short circuits
  motors, 105–106, 112–114, **113**, 129–130
  split-phase motors, 129–130
  transformers, dry-type, 61
shorted coils in motors, 112–114, **113**, 112
single row, deep-groove ball bearings, **195**, 196
single-shielded bearings, 194
sleeve-type bearings, 194, 202

spherical-roller bearings, **195**, 197
spherical-roller thrust bearings, **195**, 197–198
split-phase motors, 117–130
  bent shaft, bearings out of line, 127
  defective centrifugal switches, 128–129
  ground reversal, 129
  magnetic noise, 126
  no start, 121–122
  open circuits, 128
  overheating, 119–120
  reversed connections, 129
  short circuits, 129–130
  slow acceleration, 118
  slow speed, 118, 122–123
  stalling motor, 120–121
  starts, then quits, 122
  tight bearings, 127
  unbalanced line current, 125–126
  vibration, 124–125
  worn bearings, 127
  wrong rotation, 118
stator problems in motors, 103–104
stethoscope for testing motors, 106
switches, megaohmmeter testing, 22–23
symbols used in electrical schematics, **244**
synchroscopes, 29–30, **31**

tachometers, 30, 32
tapered-roller bearings, **195**, 198
thermometers, electrical, 33
three-phase, delta-wound motors, 140–143, **141**
  continuity testing in, 140

three-phase, delta-wound motors, (*Cont.*):
  voltage testing in, 141–143, **142**
time-domain reflectometers (TDRs), 34
transformers, dry-type, 57–64, **58**
  breakers/fuses open in, 64
  burned insulation in, 64
  conductor loss in, 63
  core loss in, 64
  distorted coil in, 63
  excess secondary voltage in, 63
  excessive cable heating in, 64
  ground fault testing in, 60–61
  grounded windings in, 61–62
  high exciting current in, 64
  high voltage to ground in, 64
  insulation failure in, 63
  open circuit condition in, 57, 59–50, **59**
  overheating in, 63
  short circuits in, 61
  smoke, 64
  troubleshooting chart for, 63
  vibration/noise in, 64
  zero voltage in, 63
transistorized stethoscope for testing motors, 106
trigger start circuits, motors, 65
triplen harmonics, 231, 233
troubleshooting techniques, 51–56
  continuity testing, 54–55
  determining symptoms, 52
  intermittent faults and, 54
  recording test results with DMM, 55
  steps in, 52

volt-amperes, 241
voltage testing, three-phase, delta-wound motors, 141–143, **142**
voltmeters, 1, **17**, 242
  ac circuit connection for, 9, **10**
  applications for, 10–11
  connections for, 9, **9**
  dc circuit connection and polarity with, 9
  fuse testing using, 16, **16**
  ground fault test using, 12–16
  low-voltage test using, 11–12
  ranges tested by, 9–10, **11**
volts, voltage, 241, 242, 252
  digital multimeters (DMM) measurement of, 41–42, **43**
  drop, voltage drop, 236–237, **237**

volts, voltage, (*Cont.*):
  Kirchhoff's voltage law, 261–265
  levels of, 227, 229–231
  low voltage testing with megohmmeter, 18–19, **19**, **20**
  low-voltage test for, 11–12
  panelboard measurement of, 229, **229**
  parallel circuits, 266–268, **267**, **268**
  sags, 230–231
  series circuits, 256–258, **257**
  solving for unknown, 261–265, **263**
  stability of, 227, 229–231
  voltmeters to measure, 8–16

watts, 241
windings in motors, 103, 105

ABOUT THE AUTHORS

John E. Traister (deceased) was involved in the electrical construction industry for more than 35 years. He wrote numerous McGraw-Hill books for electrical professionals, including *Security/Fire Alarm System Design, Installation and Maintenance, Second Edition; McGraw-Hill's Illustrated Index to the 1996 National Electrical Code; Industrial Electrical Wiring;* and *Handbook of Electrical Design Details.*

H. Brooke Stauffer is the Director of Codes and Standards for the National Electrical Code Association (NECA) in Bethesda, Maryland. He has written several books and more than 200 magazine articles about electrical and electronic technology.

JOURNEYS OF FAITH

From Canada to the World

by
Valerie J. Friesen

Nazarene Publishing House
Kansas City, Missouri

Copyright 2001
by Nazarene Publishing House

ISBN 083-411-8459

Printed in the United States of America

Editor: Wes Eby

Cover Design: Michael Walsh

Unless indicated, all Scripture quotations are from the *Holy Bible, New International Version*® (NIV®). Copyright © 1973, 1978, 1984 by International Bible Society. Used by permission of Zondervan Publishing House. All rights reserved.

Contents

Foreword		7
Acknowledgments		8
Introduction		9
A Memorial Journey A Tribute to Joyce Blair, Belize		11
1	A Jumping-Right-In Journey Elva Bates Morden, Swaziland	15
2	A Journey Step-by-Step Jean Darling, India	29
3	A Journey of Joy Hilda Moen, India	43
4	A Journey of Abundance Lenora Pease, India	57
5	A Never-Ending Journey Mary Wallace, Nicaragua and Chile	73
Pronunciation Guide		87

Valerie J. Friesen was born to Bert and Marie Friesen and raised in Red Deer, Alberta. She is the youngest of five children. Her happiest childhood and teenage memories, she recalls, were spent at the Nazarene Youth Camp in Harmattan, Alberta.

Valerie graduated from Canadian Nazarene College* in 1981. Two years later she graduated from Nazarene Theological Seminary with a master's degree in religious education (M.R.E.). She has served as a minister of Christian education in Abbotsford, British Columbia. In 1987 she was commissioned as a minister of Christian education in the Church of the Nazarene.

Valerie earned a master's degree (M.A.) in counseling from Liberty University in Virginia in 1992. She now owns and operates a counseling clinic, helping people with personal and relationship issues. She is an active member of Guildford Church of the Nazarene in Surrey, British Columbia.

---

*On January 1, 2000, became Canadian Nazarene University College.

Foreword

C. T. Studd, athlete and missionary, condensed to one sentence the grand motivation that controlled his life: "If Jesus Christ be God and died for me, then no sacrifice is too great for me to make for Him."

This book, which in a modest way briefly sketches the humble missionary service of several contemporary Canadian women, illustrates this theme.

What makes a missionary? I cannot say. It appears that God uses providential circumstances and events to mold the character of true missionaries.

In these pages, however, one clear theme emerges: At whatever cost, total commitment to Jesus Christ and doing His perceived will in full obedience becomes the secret of a full, satisfying life.

The author of these pages has a love for people and a heart for missions. Behind the anecdotes, narratives, and quotations is one simple story: God's way is always best, and in His revealed will, if we seek and wait for it, are all the ingredients for ultimate successful living.

—Arnold E. Airhart

Acknowledgments

I wish to extend thanks . . .

- To the five women whose outstanding stories are told in this book, who gave me the privilege of recording their stories and entrusting me with them.
- To Margaret Blair, who provided information about her sister, Joyce Blair.
- To Clayton and Noreen Mills, lifelong friends, who provided invaluable advice in the early stages of working on the manuscript.
- To the pastor, members, and friends of Red Deer First Church of the Nazarene, who taught me to love God and the church.
- To my parents, Bert and Marie Friesen, who taught me about the excitement of missions and, through their example, the satisfaction of helping others.

Introduction

Incredible stories. Incredible journeys. Incredible faith. These women were born and raised in Canada and spent their lives overseas, serving the Church of the Nazarene in mission service. The accounts of these individuals describe their Canadian heritage and the ingredients that shaped their lives. Details of their God-whispered calls, missionary preparation, daily challenges, and work assignments are chronicled here.

These women were not among the Nazarene mission pioneers of the early 20th century. Rather, they are second- and third-generation missionaries, each in her own way contributing to the initial work that was started. They candidly tell about the tough times and how they coped. They share their joys and rewards. They give us a glimpse of what a typical day on the mission field is like. And from their experiences we learn timeless truths that will encourage and direct us as we proceed on our own journeys.

Canada has been blessed with many trailblazing women who have planted Nazarene churches across the great dominion. These women in this collection were trailblazers of their own kind, instrumental in the work of the Church of the Nazarene in British Honduras (Belize), Swaziland, India, Nicaragua, and Chile. Though only representative of the women from Canada who have served as Nazarene missionaries, these ladies leave a wonderful heritage for all of us to follow, and this book is a record of the legacy.

A Memorial Journey

A Tribute to
Joyce Blair, Belize

Possessive? Most definitely! Unashamedly possessive!

The people of the First Church of the Nazarene in Red Deer, Alberta, "owned" Joyce Blair. She was *their missionary,* and the church members were *her folks.* Respect and love were mutual.

Joyce Blair

Joyce became a Nazarene at Red Deer First. She and her family worshiped there for many years, and she left for the mission field from the church. Joyce Blair was a common household name among Red Deer Nazarenes. They faithfully and fervently prayed for her throughout her 33 years as a missionary. They pampered her during times of furlough. They honored her by naming a room in the Christian education wing the "Joyce Blair Missionary Room" and a mission study group the "Joyce Blair Missionary Chapter." And in retirement, they warmly welcomed her again, making her a vital part of the church fellowship.

Joyce, the oldest of eight children, was born to Sandy and Bertha Blair on November 18, 1911, in Admiral, Saskatchewan. Her parents christened her Annie Mary Joyce Blair, but she was best known by her family and friends as Joyce.

When she was seven, the Blair family moved to a farm southeast of Red Deer. Her father operated a small post office while managing the farm. For grades one through nine, Joyce attended a one-room school near her home. For high school, she boarded in Red Deer, where she was introduced to the people called Nazarenes.

Joyce graduated as a registered nurse from the Royal Alexandra Hospital in Edmonton in 1933. She acquired practical experience at Parson's Clinic in Red Deer. Later, she continued her studies at Pasadena College (now Point Loma Nazarene University) in California, earning a bachelor of science in nursing.

In 1943 the Church of the Nazarene appointed Joyce as a missionary nurse to the country of British Honduras, now Belize,* in Central America. When she arrived in Benque Viejo del Carmen, there was no doctor within 50 miles. She had the sole responsibility for delivering babies and often traveled into the jungle at night to visit patients. In addition to opening a clinic, she trained nurses, taught in the Bible college, and preached in the churches.

A special aspect of Joyce's missionary career was her dedication to three homeless boys. She provided the lads with a home and education, becoming their surrogate parent and confidant. Sadly, one of the young men died at age 26; however, the other two are married and living in Miami.

On one of Joyce's furloughs, she was ordained as a deaconess on the Canada West District.

After more than three decades as a Nazarene missionary, the entire time in Belize, Joyce retired to her home in Red Deer. For a few years, she worked at the Red Deer Nursing Home, where later she became a resident. Many summers she was the nurse at the Harmattan Nazarene Youth Camp, where she lodged in a rustic log cabin, heated with only a potbellied stove.

"Miss Joyce," as she was affectionately known, died on November 12, 1988. Beloved and esteemed by old and young, Joyce Blair made a profound, indelible impact on everyone her life touched, whether in Red Deer or Belize. Her life is intricately intertwined with some of the women whose stories are related in this book. Thus, how fitting this memorial journey!

---

*A guide on pages 87-88 is provided to help in pronouncing unfamiliar words.

A Jumping-Right-In Journey
Elva Bates Morden, Swaziland

"Jump right in" and "let me try." These expressions epitomize the journey of faith for Elva Bates. When others would say, "I can't do that," Elva would jump right in and do what needed to be done. She would tackle any task with the attitude "Let me try, and if it doesn't work, then I'll try something else."

Elva Bates Morden

Jobs were plentiful on the mission field. Supervising clinics, instructing nurses, dispensing medicines, inoculating children. Teaching Sunday School, training teachers, preaching sermons, planting churches. Repairing furniture, fixing broken windows, replacing screen doors, patching roofs. Elva did not think about whether she was able to do the tasks. She just did them. Her journey of faith included countless jumping-right-in moments.

Elva Bates accepted a missionary appointment by the Church of the Nazarene to be a nurse in Swaziland in southern Africa. But her duties extended beyond medical responsibilities to include church-related work and handywoman tasks.

In reflecting upon her early years on the mission field, Elva explains: "I was directly involved with community health issues and practical health treatment, responsible for supervising several clinics in rural areas. Central to my work was the organization of child welfare and antenatal [prenatal] programs in all of these clinics."

When she first arrived in Swaziland in 1961, children who lived to the age of eight would probably make it to adulthood. Even though progress had been made through Nazarene medical missions, the mortality rate below eight was very high. It was not unusual to have a woman come into the child welfare clinic or an antenatal clinic, pregnant with her 13th or 14th child. She might have one or two children living; the rest would have died before their fifth birthday. Over the years Elva and her contemporaries labored in Swaziland, the mortality rate improved significantly. If children lived until age two, they would likely then live full, healthy lives.

Helping the people of Swaziland with family planning was difficult. Because of the number of deaths in a family, Swazis continued having babies so there would be several grown children to care for them in their old age. But what the Swazi people did not realize was the fact that additional children would affect the health and education of the ones they already had. Malnutrition and education were significant concerns, as the resources of the Swazi parent could only stretch so far.

Nurse Bates at Kashewala Clinic

Immunization was a major part of community health. Children were weighed, checked, measured, and inoculated on a regular basis. They returned home from the clinics with a little milk to help improve nutrition. In addition, nurses provided instruction about hygiene, breast feeding, and formula preparation.

Development of the clinics did not happen overnight. Each one was established only after much hard work. When Elva arrived in Swaziland, the 13 existing clinics, started by her predecessors, were outposts or extensions of the Raleigh Fitkin Memorial Hospital in Manzini.

Eventually, the nationals from the Nazarene Nursing College were trained to handle the local clinics. Through the educational process, nurses were trained to work alongside missionaries and then manage the clinics independently.

Doctors usually visited each clinic one day each month. For example, Dr. David Falk, a Canadian missionary, had an intense interest in community health and frequently worked at the clinics. Most of the time, however, health care at the clinics was the responsibility of the nurses. Often, 100 or more patients would visit each clinic each day.

Patient records were kept on small cards. While the Swazi people stood in line, according to the order they arrived at the clinic, as much history as possible was collected. On the days both men and women came for treatment, males were seen first, as the one-room clinic could not accommodate both genders simultaneously.

Most clinics, such as Piggs Peak in northern Swaziland, were located in remote areas, and often Elva would be the only missionary available. It was not unusual to be 70 miles from the hospital and 13 miles from the closest mission station. Despite the distances, Elva managed well and was able to deal with each problem. During these times, Elva's jumping-in-and-doing-whatever-needed-to-be-done attitude had an enormous impact. Unquestionably, God's preparation of Elva for missionary service paid off—time after time after time.

🍁

> Quickly learning to fend
> for herself, Elva became
> quite independent.

🍁

Journey Through Childhood

Elva was born on June 23, 1931, to proud parents Alymer and Pearl Bates in Prince Albert, a city in northern Saskatchewan. She had two siblings: an older brother, Don, and a younger brother, Doug.

The "zillion" boys in Elva's neighborhood shaped her into becoming a tomboy, playing hockey, baseball, and other guylike activities. Being part of the group was a priority. Quickly learning to fend for herself, Elva became quite independent, enabling her to figure things out on her own. She did not hesitate to tackle something typically considered masculine. Seldom was it necessary to seek assistance. This neighborhood experience helped develop necessary coping and problem-solving skills that she would use many times throughout her life.

Looking back, Elva believes her mission call was being worked out throughout her childhood. After hearing that she had been appointed to the mission field, she telephoned a friend to tell her the news. The friend was not at all surprised and reminded Elva of the time when they were in the third grade. Their teacher, walking up and down the aisles, asked the students what they were going to be when they grew up. Standing up, Elva stood straight and tall, shoulders squared, and declared, "I'm going to be a missionary."

The Journey with God

The Bates family faithfully attended the Presbyterian church. Here Elva established a good foundation in the Bible; however, she did not have a clear understanding of salvation. This doctrine had neither been taught in Sunday School nor preached from the pulpit.

Elva heard about salvation from a friend. Through this friendship she learned of the importance of having a personal relationship with the Lord. Another friend invited her to the Church of the Nazarene, where the gospel was preached, taught, and modeled. What she heard and witnessed was much like the Scriptures she had been reading. All along Elva had been searching, and she found the answers to her questions at the Church of the Nazarene. She accepted Christ as her personal Savior on January 11, 1948. Elva's journey of faith began.

Elva confesses that one of the greatest disciplines of her life was her involvement in church. Besides regular attendance in church and Sunday School, she loved summer youth camps and the CGIT (Canadian Girls In Training).

"When I was 17 years of age," Elva testifies, "I received a specific call to the mission field during a missionary's message. At the end of the sermon, there was a call for commitment of young people to serve the Lord. I went forward that night and said yes to God." This was the start of her missionary journey.

The Education Journey

Elva knew that a professional career of some type was necessary for appointment to the mission field. She considered both teaching and nursing, the two main choices for women in the 1940s. Since she preferred being a nurse, she chose that profession. She also figured if she was turned down for this vocation, she could always apply for teaching.

As it turned out, straight out of high school, Elva enrolled in nurses' training at the Holy Family Hospital in Prince Albert, her hometown. She developed several close friends during these years. Some of them, including Hilda Moen, Barbara Cogger Blair, and Loreen Amonson Fleming, were a part of the youth group at Prince Albert Church of the Nazarene. At least five from that group went out as missionaries, though not all with the Church of the Nazarene. Those Nazarene youth had a lasting impact on Elva.

Following nurses' training, Elva moved to Red Deer, Alberta, to work in the local hospital for one year to save money to attend Canadian Nazarene College (CNC),* which was located in the same city. After two years of study at CNC, she moved to Wainwright, Alberta, to nurse once again and save money for two more years of college. Elva knew that she probably needed nursing experience and theological education to be considered as a mission candidate.

Elva returned to Red Deer to complete her program at CNC. She felt she should know more about what she believed and why. Desiring a solid base in the Bible and church doctrines, she felt a Nazarene college would be the best way to accomplish this goal. Elva did not know all that she would be doing on the mission field. If she had to preach, then she wanted to be prepared. CNC also brought lifelong friends into Elva's life—Willa Witte from British Columbia, Joan Oliphant from Ontario, and Joyce Slang.

The Missionary Application Journey

The process of becoming a missionary happens over many years. First, there is the spiritual process whereby God calls people for mission service, im-

---

*On January 1, 2000, became Canadian Nazarene University College.

printing the strong desire on their minds and hearts to serve Him. Second, there is the formal application process that missionaries with the Church of the Nazarene must engage in.

Elva sent the first application during her second year of college. When the director of the Department of Foreign Missions (now World Mission Division) in Kansas City visited CNC, Elva had an opportunity to talk to him face-to-face. Each year, Elva wrote to the department, reiterating her continued interest.

Later, Elva submitted the second required application along with the completed medical examination. In 1960 Elva was invited to Kansas City, which meant she would meet the General Board. A few weeks following this trip the hoped-for letter arrived. Elva had been appointed as a missionary for the Church of the Nazarene.

"When word eventually came that I was appointed to Swaziland, the news was beyond comprehension," Elva recalls. "I felt as if I was walking with my feet two feet off the floor. Finally, things had come together, and I was on my way to Africa."

The Swaziland Journey

I'm here at last! Elva thought upon arriving on the mission field in 1960. Her eyes moved about, drinking in the African landscape and her new home. "Thank You, Lord," she prayed silently. "Thank You for allowing me to be Your servant here in this beautiful country."

In Endzingeni, a small village in northern Swaziland, Elva taught Sunday School and helped train the teachers. One of the obstacles was the lack of teaching materials and reference books. Tackling this problem with her let's-do-it spirit, Elva met with all of the teachers each Thursday afternoon to teach them the lesson, demonstrating techniques to be used the next Sunday morning.

Elva Bates working with Sunday School children

Elva also served as pastor and church planter. During the last two terms, Joyce Vilakati, her Swazi coworker, and Elva established a church in a witch doctor's school. Unashamed, they approached the old man to see if they could have services in the school building. Not only were they given permission, but also eventually the witch doctor and his wife, plus several of the children, were saved. Soon after, the witch doctor school closed its door, and the church flourished.

Elva used the same model for ministry that she used in nursing. "I always aimed at working projects in such a way that when I left, they would continue on without skipping a beat," Elva says. "I did not want the nursing or church work to depend entirely on me." This proven method emphasized the importance of training others and helping them become leaders. It took more work, of course, but Elva didn't mind. The results for Elva were rewarding, and for Swazi Nazarenes, long-lasting and effective.

Elva dispensing medication at the Piggs Peak clinic

Elva's mission service was a thrill and, at the same time, a curious challenge, for she never knew what she was going to find. The sheer pleasure and extensive demands of her work compelled her to do far more than her training and skills. She knew that, most of the time, no one else was available. So her jump-right-in-and-do-it temperament prevailed.

Part of the supervision of the medical clinics was ensuring adequate supplies. When Elva first walked into the pharmacy, she stared at the shelves of medicines and discovered that the labels were unfamiliar. Elva had to pull out reference books to learn the strange names. Although many drugs were out-of-date to Elva, they were still effective in treating the people. The Swazis had not received as much medication as people in more developed countries; therefore, their bodies responded to the medicine just fine.

🍁

> To this longtime single missionary, Clarence's marriage proposal was rather surprising.

🍁

Journey Down the Aisle of Matrimony

Me? A blushing bride? At my age? Elva talked to herself as she studied the note in her hand. To this longtime single missionary, Clarence's marriage proposal was rather surprising. It all began this way.

On Elva's second furlough in 1972, while touring the Canada Pacific District (British Columbia and Yukon), she stayed in the home of Clarence and Kay Morden in Nanaimo, B.C. She developed a strong friendship with them and corresponded with Kay regularly.

When Kay became ill with cancer and unable to correspond any longer, Clarence promised to write Elva and other friends to let them know what had happened. Following Kay's death, Elva sent Clarence an expression of her sympathy. Twice a year, at Christmas and on her birthday, Clarence wrote Elva, enclosing a small present. Elva always responded with a thank-you note. The semiannual correspondence between Canada and Swaziland continued for many years.

Since Elva had no definite plans for retirement, she was not sure how she would care for her aging mom or how she would manage financially. Yet, amid this uncertainty, God was already planning her future.

When Elva became ill in Africa, she wrote the World Mission Division and requested permission to remain home after her next furlough. Right after sending this letter, the unexpected offer came from Clarence. "This proposal by mail was so exciting for me," Elva remembers. "Yet, my mind was whirling, for there were so many things to think about."

Elva confided in some missionary friends, Bert and Marie Friesen, about the marriage proposal. Once Elva decided, then Bert suggested that she telephone Clarence to let him know of her acceptance, rather than writing him a letter, as her long-distance suitor had already been waiting several weeks for a response.

Elva thought that she should wait until she finished her term, which meant she would stay in Swaziland for several more months to fulfill her commitment to the mission. Elva also decided that she would tell her mother personally. Clarence was sworn to secrecy; he could not tell anybody until Elva had told her mom and the World Mission Division.

Everything worked out better than Elva could have dreamed. When she arrived home, she informed her mom, notified church headquarters, and soon the wedding plans were under way. The churches in Prince Albert, Saskatchewan; Lacombe, Alberta; and Nanaimo, British Columbia, all announced the engagement of Elva Bates and Clarence Morden the same Sunday. It was now public!

Elva completed more than 23 years of missionary service when she retired in February 1985. A month later, Elva married Clarence Morden in the Church of the Nazarene in Lacombe. For several years they lived in a spacious house, overlooking the beautiful Pacific Ocean at Lantzville on Vancouver Island. In 1999 Elva and Clarence moved to Abbotsford, British Columbia.

"I never felt that my life was a sacrifice," Elva concludes in speaking of her missionary journey, "for I have always done what the Lord wanted me to do. It always seemed as though everything just fit. Even when I had to jump right in. Praise the Lord!" ♣

A Journey Step-by-Step

Jean Darling, India

No classroom. No curriculum. No instructors. No charts. No students. No anatomy models. No equipment. And, to top it all off, a language barrier. Yet, Jean Darling's assignment was to start a training school to produce skilled nurses to work alongside the doctors at the Reynolds Memorial Hospital in Washim, India.

Jean Darling

Just imagine sterilizing instruments and syringes by heating water in an enamel kettle over an open woodstove. Sterile gloves are lacking. The hospital staff are accustomed to brass and aluminum rather than the brand-new glass thermometers and syringes. The greatest obstacle is the strong negative public perception of the nursing profession.

Up stepped Jean Darling, a 27-year-old. The assignment was daunting. How could she begin to put together all that was needed? Where would the equipment come from? Did she personally have what it would take to get the job done? Where should she start? What brought Jean to India in the first place?

Jean's Missionary Call

Jean recognized that God planted seeds of thought and desire and through the years let them grow. Later, He confirmed that she was indeed exactly in the place where He wanted her to be.

"For as long as I could remember," Jean says, "my desire was to be a missionary in India. I was not sure where the idea came from, but I found out that my grandmother wished to be a missionary, even though she never was able to go. I was only two and a half years old when she died. Yet in the back of my mind, I have often wondered if the idea came from her."

Jean completed high school before becoming a Christian. She took nurses' training in London, Ontario, at Victoria Hospital. In 1939 after one year of education, at the invitation of a classmate, Jean attended a revival meeting. That night she gave her heart to the Lord.

After her conversion, Jean thought much about India. Had this been a call from God or just her own idea? She wondered if the Lord would ever call anyone before becoming a born-again Christian. Jean decided to cover all bases by attending Bible college. If it was a call, she would be ready; if not, the theological study would be good for her anyway. Yet her struggle continued. One night she prayed for hours. Finally, God made it plain that being a missionary was His call and will, that He would make the way for her to live and work in India.

God further confirmed the missionary call to India when Jean met the General Board of the Church of the Nazarene. "I literally shook in my boots for fear I would not be accepted," Jean remembers.

Jean, riveted to her seat, listened to Mary's account of conditions in India.

After the meeting with this august body, she had an interview with a second group, the Board of General Superintendents. "Are you willing to go to any field to which you may be assigned?" one of these esteemed leaders asked.

"I was stunned," Jean admits. "I thought I had a clear call to India. I tried to calmly indicate my willingness to go anywhere. Inside, I knew that God never makes mistakes; so, I left the door open for any possibility." Jean accepted this new development with a profound inner assurance. Later, with no fanfare Jean learned that she had been appointed to India.

God was not yet finished with assuring Jean that India was where He wanted her. Mary Anderson, a missionary with the Church of the Nazarene in India, spoke at the church in London, Ontario. Jean, riveted to her seat, listened to Mary's account of conditions in India. Jean recalls that Mary's description of the hospital, an old mud school building with minimal equipment, had helped prepare her for what India would be like. Now with more realistic expectations, Jean adopted the attitude that she had to be ready for anything.

Language study proved to be both interesting and rewarding for Jean, as language was one of her best subjects in high school. Marathi, one of the many languages spoken in India, was considered to be a difficult language for English speakers to learn. Jean passed the written as well as oral exams for the first and second year of study, a requirement before full-time mission work. Jean felt certain that God had been carefully preparing the way for her work in India, even through language study. Before she was done, she could fluently pray in Marathi, even before she could speak it well. "What a joy to be able to pray with and for my patients in another language, their mother tongue," Jean recalls. It was just one more affirmation that God planned for Jean to be in India. The seed of working in India had grown, coming to full bloom.

Jean's Childhood and Education

Jean was born on October 26, 1918. Her family lived on a farm 20 miles northwest of London, Ontario, near the small village of Clandeboye. Her mother, Emily Ada Whitford, who had a Scotch-Irish background, was raised in the Church of God (Anderson, Indiana) with a Holiness tradition. Her father, William, whose parents were Methodists with Scotch-English background, did not attend church.

"I was the fifth child in a family of eight," Jean says. "My siblings were three older brothers; an older sister, Grace; two younger brothers; and a younger sister, Audrey. We always said that there were enough of us to have our own baseball game. During high school, Grace and I played in a softball league, and we thought nothing of running two and a half miles to catch a ride to the games. I loved to play softball when I was not needed to help with farm chores, such as weeding the large family garden."

When Jean was ready for college in 1943, after she had worked for two years as a nurse, Jean chose to attend Eastern Nazarene College (ENC) in the United States. Her hometown was part of a Canadian district included in the ENC region. She learned nurses were not allowed to cross the border during World War II; so, she applied to Canadian Nazarene College in Red Deer, Alberta, attending there from 1943 to 1945.

Canadian Nazarene College was a perfect campus for Jean, as she was shy and thought she might have been lost in the crowd at a larger school. Upon graduating in 1945 from the two-year Bible course, she began preparing to leave for India. After packing and sending heavy luggage by ship via New York, Jean headed in the opposite direction for San Francisco.

Jean's Trip to India

Traveling at the end of World War II was adventurous, which made Jean's travel to India all the more exciting, yet difficult. Since the crowded trains had no berths, Jean had to sit in a regular seat for four days and four nights. Upon arrival in San Francisco, a pastor was to meet her, but he did not get the message. With her purse for a pillow, Jean slept in the railway station. Every few hours she called the minister's home. Eventually he answered the telephone. He came and took her to the hotel, where she joined Ruth Freeman, who would travel with her to India.

The next leg of the trip was by ocean to Brisbane, Australia. Jean and Ruth booked passage on a warship with triple-decker bunk beds for sleeping accommodations. On the same vessel were other Canadian women who had married Australian men during the war and were traveling to join their husbands. By mistake, the luggage for all the Canadians was stored in the ship's hold and thus inaccessible. Jean had no change of clothing for the entire trip. Some relief came when a Russian movie star loaned her a jumper, even though the hem hung at her ankles. Fortunately, at the end of their journey the luggage was retrieved and returned to the passengers.

In Brisbane, the young missionaries went by train down the coast to Sydney, where Jean and Ruth met 19 other missionaries on their way to India too. However, the adventure was not over; there was still one other hitch. Due to the war's end, letters from the travel agency in New York had not arrived in Sydney. None of the 21 missionaries had tickets to get them to India. There was a ship in the harbor bound for India, but it was under repair and all booked as well.

"Ruth Freeman and I were staying in a room close to the shipping office," Jean recalls, "and we inquired daily about a passage to India. One day, Ruth and I learned that there had been some cancellations, which would make it possible for us to buy tickets. We telephoned the other missionaries to inform them that there were enough cancellations to take care of all 21 of us.

"Finally, I arrived in Madras, India, on December 31, 1945, after a three-month journey. A missionary, Prescott Beals, came to meet me and accompanied me by train to Malkapur. From Malkapur we motored to Chikhli. It was so good to be 'home'—at last."

Jean's First Missionary Assignment

Jean soon encountered the daunting job of organizing the nurses' training school. She quickly realized her strongest asset was an unwavering faith in God, which gave her the confidence and strength to face the new challenges. She knew that she needed an unusual creativity, practical ingenuity, and gritty determination to solve the many problems. And God granted all these.

Plans began to take shape. A grassy courtyard for sunny, dry days and the outpatient department for rainy, wet days were selected for the classroom. Later, beds and tables were brought in for demonstrations and desks. A life-sized doll, sewn from cloth and mattress stuffing, was used as a teaching model. Jean wrote the curriculum in English, and a local pastor translated it into Marathi. She purchased charts and collected bones to shape into skeletonlike figures. Eventually, a large pressure

Jean Darling as nursing instructor

cooker became an autoclave, and sterilization techniques were taught.

The first class of four students learned obstetrical, pediatric, medical, and outpatient information. On one occasion, Jean left the students on their own to practice a technique. When she returned, Jean found the students working with patients down on the floor. Student nurses were not accustomed to working with beds or tables, and patients preferred to sleep on the floor.

The students learned they could trust Jean to shape their reputation as nurses.

Jean had to establish the confidence and respect for the nursing profession among the people. One time a young man, after carefully watching Jean administer typhoid shots and perform her nursing duties, brought his entire family to receive their vaccinations. By observing Jean, he had been convinced that she knew what she was doing. Another time when soiled dressings fell on the floor, the students were surprised when their teacher bent over to pick them up. Usually this would be a task only a sweeper would do. Gradually, by Jean's example, the students learned that it was OK to pick up bandages and they could trust Jean to shape their reputation as nurses.

Others came to help at the hospital and nurses' training school. They taught classes, collected more equipment, and developed curriculum. The school began to grow. In spite of all the obstacles, sometimes discouraging and disheartening, Jean was able to work alongside others and fulfill the most challenging of assignments.

Jean was persistent and steady. She did not attempt to turn the world upside down in a day. Most of the time she was known to be easygoing, although she sometimes apologized for her impatience when others failed to give attention to details. A student once described her, "Sister is very strict, but she understands us and loves us." What a great compliment!

Jean's first class graduated with local hospital certificates in 1949. Four years later India's government recognized the nurses' training school. This accreditation was the pinnacle of the nursing program. Furthermore, the nursing students had gained respect, recognition, and acceptance as professionals. Today the responsibility for medical work at the hospital is in the hands of capable Indian leaders with a full complement of nurses.

When Jean was being trained as a nurse in Canada, she was taught to be conservative in her professional dress. She never wore slacks, always wearing a dress and nylons. Upon her arrival in India, she wore dresses, hose, and sandals. One day a lad came up, pulled her nylons, and exclaimed, "Mommy, Mommy, her skin is loose!" Jean quickly realized that stockings did not need to be a necessary part of her attire. Besides, nylons were just too hot.

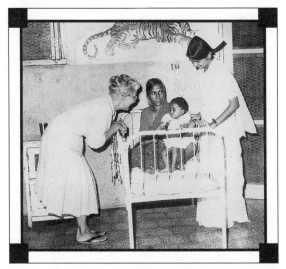

Jean on duty in the pediatrics ward

Jean's Business Office

Jean had been concerned about mission finances and needed someone to organize the business office at the hospital. One day while praying, she thought, *Why can't I organize the finances?* The next day, she received a letter in the mail with the poem "Step-by-Step," which spoke directly to her. The first verse said, "Don't worry about what you do not know because I already know and will lead you *step-by-step*" (Jean's paraphrase). That is exactly how it turned out.

Whenever Jean visited with Hilda Lee Cox, Bill Pease, or someone from another mission, she usually heard them talking about the very things she needed to know. To her, this was an answer to prayer. In addition, a young Nazarene lady, a recent high school graduate, came to help.

While Jean was on furlough in 1960-61, missionary John Willis Anderson took charge of the business office and hired a young man, Eby Bansod. From Eby's previous experience as a schoolteacher for two years, he could speak English and type. What a find! Eby made a great addition to the office staff, and with his help the business office became more efficient. Jean recalls, "It was an exciting day for me when Eby was appointed the business manager and I became his assistant."

After returning to India in 1966 following her next furlough, Jean was assigned to the boarding school and dispensary at Chikhli. There was no one to replace Canadian missionaries Bill and Lenora Pease, so Jean took over their jobs at the school as well as the work of the treasurer. In 1981 Jean returned to Chikhli to replace Hilda Moen, also a Canadian missionary, who went on furlough. Jean worked at the school and clinic until she retired in 1985.

"As I look back, I was keenly aware how everything I did seemed to prepare me for the next assignment or task," Jean recalls. "The business office experience at the hospital prepared me for the work at the school as well as the treasurer's job. The mission treasurer was automatically chairperson of the auditing committee, which gave me insight into the local churches. This in turn prepared me for the assignment of mission chairperson [director] in 1973."

Jean's Friend Bhakshibai

Among Jean's many friends was Bhakshibai, whose conversion story began when her husband was a patient in the hospital. His private room was near the chapel. Since all patients and relatives were welcomed at the morning chapel service, Bhakshibai came and sat on the floor just inside the door. Above her head hung a picture of Jesus surrounded by children. Right next to Jesus was a little Indian girl in a sari. Bhakshibai found it amazing that, of all the children representing different countries, an Indian girl was next to Jesus. It was through that picture that Bhakshibai came to know that Jesus loves everyone, even little girls.

Bhakshibai

Soon after, Bhakshibai asked Jesus Christ to come into her life. She brought her daughters-in-law and grandchildren to the hospital. Bhakshibai also took medicines and directions home to administer to her family.

"How do you give the medicines to the different members of her family?" Jean asked.

"I get the medicine ready," Bhakshibai replied. "Then I sit in front of the patient and say, 'I give you this medicine in the name of Jesus.'" Bhakshibai was convinced that God was blessing the medicine and making her family well.

Jean's Retirement

After 40 years as a missionary in India, Nurse Darling retired in 1985. She lives in London, Ontario, in the same complex as her good friend and younger sister, Audrey. "When I reflect on my years in India," Jean says, "phrases like 'my walk with God,' 'not I but Christ,' and 'Holy Spirit, be my Guide' come to mind. There were many days of just plain, hard work. There were even those days of darkness and discouragement, when it seemed as if I could not see ahead. Then God would come to lighten my path and show me the way. I always remembered those timely words from the poem 'Step-by-Step'—'I already know and will lead you.' The Lord has led me, step-by-step-by-step." ✤

A Journey of Joy
Hilda Moen, India

Hilda sat glued to her seat, her hands clasping the hymnal. Completely absorbed, she listened to the speaker, Louise Robinson Chapman. The need for missionaries and how God asks individuals to serve Him on the mission field were carefully detailed in the message.

Hilda Moen

Then Hilda's mind drifted. "All I could visualize were little children sitting on the floor and holding out their hands, although I did not have a clear idea of what is meant to be a missionary," she admits.

At the close of midweek service, Mrs. Edward Lawlor sang "Victory in Jesus" and gave an invitation to come to the altar. Hilda stood to her feet and walked to the altar without hesitation. Without fully knowing what God had in store for her, yet with calmness, she went forward that night to say yes to the work of the Lord through missionary service. Hilda's journey of joy had begun.

The many years spent growing up on the family farm would certainly be an asset for overseas mission service. The work, abundant and difficult, involved numerous chores. Cows to milk. Eggs to gather. Water to carry. As a teen, Hilda spent long hours in the fields driving horses, discing, ploughing, cultivating, harrowing.

Endorsement of her decision to be a nurse and missionary by Hilda's godly parents provided significant assurance. Her folks, quiet, unassuming people of Norwegian descent, did all they could do to adequately prepare their daughter for the challenges she would encounter. Her mother spent many hours kneeling at Hilda's bedside, praying for her.

In spite of the factors that ideally shaped Hilda's life for mission service, she had other more practical concerns of living in a foreign country. Fear of snakes and disease—malaria, smallpox, and typhoid fever—were uppermost in her mind. "Furthermore, one of the greatest problems I had," Hilda confesses, "was the feeling of being incapable of doing the job. I could not sing. I could not play the piano. I had little experience in public speaking. And I was unsure of how I would learn to speak a new language, let alone having others understand me."

But Hilda quickly learned that a willing-to-go-at-any-cost resolve was the most important qualification and asset. Then God would do the rest.

Formal education consisted of one year at a Lutheran Bible college in Outlook, Saskatchewan. Then Hilda began two years of nurses' training at the Holy Family Hospital in Prince Albert. After graduating in 1950 as a registered nurse, she moved to Red Deer, Alberta, to attend Canadian Nazarene College (CNC).

Telling Bible stories in a creative, attention-grabbing fashion became Hilda's forte.

While attending CNC, Hilda did double-duty work, nursing and studying. In addition, she cultivated the skills of discipline, time management, and financial responsibility. She also learned how to live on little sleep. All of these were necessary ingredients for the demands that would be placed upon her in the future. After four challenging years, Hilda graduated with a Th.B. (bachelor of theology) in 1954.

Practical experience, which began in Prince Albert, was expanded and enhanced in Red Deer and later in Swift Current, Saskatchewan. While in the latter town, Hilda served as associate pastor, greatly involved with visitation, preaching, and even administration. Here she developed competencies she would need again and again in helping people on the mission field.

Telling Bible stories in a creative, attention-grabbing fashion became Hilda's forte. Mina, her sister, who became a missionary to Ethiopia, taught Hilda the basics. As an example, Mina told the story of Fluffy, the lost sheep. Then Hilda used the same story with her Sunday School class. The principles of storytelling proved to be invaluable to Hilda, whether relating stories in India or preaching deputation sermons in Indiana.

The Joyful Appointment

Hilda was called to meet the General Board and the general superintendents for an interview in January 1956. What a thrilling, joyful journey! She was reassured by the presence of Louise Chapman, who was on the General Board at that time. Dr. Edward Lawlor, also a board member, had previously been her district superintendent. Both he and Mrs. Chapman attested to Hilda's life, testimony, and missionary call.

The interview process went smoothly, and Hilda was appointed as a missionary. Plans took shape quickly, and that October the new missionary was on her way to India. The period of preparation was over. The time had come to begin her work.

The Joyful and Not-So-Joyful Arrival

Hilda rendezvoused with Wallace and Phyllis Helm in Hawaii and traveled with them to India. Missionaries Earl and Hazel Lee, Geraldine Chappell, and Juanita James greeted them at the airport with fragrant garlands. The heat was scorching; the sojourners, exhausted. After a lengthy examination and clearing customs, the new missionaries took a taxi to their hotel rooms. This initial trip gave Hilda her first view of the countryside. Transfixed by the hovels that lined the road, she began to grasp the dire needs of the people she came to serve. "This is India. I had finally arrived," Hilda remembers thinking. Then she adds, "I slept well that night." Her journey of joy had taken her halfway around the globe.

The next day began with the business of living in a new country. The first agenda item was to open bank accounts. The Canadian currency had to be changed into rupees and annas. Money, Hilda soon realized, would be one of numerous challenges. She had so much to learn.

Outside the bank and around the corner, a crowd gathered on the street. Hilda moved closer to investigate the ruckus. Peering over and around the people, she saw a dark-skinned, elderly man with a large wicker basket. Lifting the basket's lid, he began to play a flutelike instrument. A king cobra lifted its head above the rim and, swaying back and forth to the music, rose higher and higher. Frightened, Hilda really, *really* wanted to dash away. Her journey had just lost some of its joy! In taking another glance, she noted that the man had stopped playing and the snake had lowered itself back into its basket home. Hilda took a big breath, sighing with relief.

Hilda soon discovered that some of her greatest fears were reality. Snakes. Rats. Scorpions. Centipedes. Spiders. And she learned these insects, rodents, and other "despicable creatures" could be found in her home. Since the mosquito is the primary carrier of malaria, a mosquito net was hung over her bed as a precaution. Even with the negatives, Hilda quickly and easily adjusted to a different culture and lifestyle.

Hilda journeyed by rail to the interior of India to her first assignment at Washim. The train had compartments where men, women, and children all slept in the same area. There were eight bunks, four upper and four lower. The toilet comprised a hole in the floor and a spout for running water to wash their hands. She was cautioned, of course, not to drink the water.

Hilda and her companions brought their "beds" with them—bedrolls of canvaslike coverings with small mattresses, pillows, and blankets. All people slept in their clothes. To climb to the upper bunk, Hilda scrambled up the best she could, as she could not locate a ladder. Though excited about the trip, she spent a sleepless night, afraid she might roll off the bunk as the train lurched from side to side. She arrived in Washim without her fears being realized.

Joy in Mission Service

A huge hurdle to overcome was mastering Marathi. A pundit, or teacher, came daily to provide instruction. Hilda first learned how to write her name. Slowly, she began to speak and converse

Hilda at the Buldana mission station

in this unique tongue. During the time of language study, she also worked part-time in the Reynolds Memorial Hospital. This provided the opportunity to practice the language as she supervised the national nurses in medical procedures. Eventually, she attended a Marathi language school in the mountains.

Hilda returned to full-time work at the hospital. The needs were certainly plentiful. All the beds were full, and more patients lay on the floor. It seemed as if people were everywhere. Outside the wards were small cookrooms where relatives prepared meals for the patients.

Later, Hilda was assigned to the health clinics, which involved moving first to Chikhli and then to Pusad. In the latter community, she worked with Dr. Evelyn Witthoff, and together they traveled from village to village, holding clinics. The doctor often treated 150 to 200 patients in one day. Hilda observed Dr. Witthoff's ability to proficiently diagnose and treat patients and frequently witnessed—with joy—miraculous recoveries.

Joy in Healing the Sick

One day at the Pusad dispensary, a sick child was brought in for treatment. The family had already taken the young girl to the village doctor, who had burned the child's chest in several places to drive the evil spirits out of the body. Infection had developed on those burned areas. As a last resort, the family brought the child to the clinic.

After a quick examination, Hilda established

that the child's lungs were full of fluid, meaning pneumonia. The only treatment that would help at that time was a penicillin injection, but Hilda was concerned about giving penicillin. She had seen a man die in the Red Deer hospital from a reaction to the drug. Hilda prayed and gave the injection. Immediately her worst fear came true: the girl stopped breathing. Hilda gave the youngster a medicine to counter the allergic reaction and started artificial respiration.

🍁

> **The mother rushed around the examining table, threw herself on the floor, and kissed Hilda's feet.**

🍁

Standing close by was Anusayabai, a medical assistant, praying. "O God, please take care of this child," she pled. "You know we need Sister Moen to stay here and help us, so please, don't let this girl die." Then Anusayabai looked up, raised her hands, and with a smile on her lips uttered gratefully, "Thank You, Jesus."

Suddenly, the child gasped and began to cry. The mother rushed around the examining table, threw herself on the floor, and kissed Hilda's feet. The startled missionary took a step back. "No! No! Don't bow down at my feet, or before any person," she protested. "Bow only at Jesus' feet."

Hilda helped the woman up, and the appreciative mother exclaimed, "But you are God! The lady was praying to God, and when I looked to see whom she was praying to, I couldn't see anyone else but you. So she must have been praying to you. You answered her prayers."

Hilda explained to the mother that Anusayabai was praying to Jesus, the living God, not a god of stone. "This God is in the heavens, and we cannot see Him with our eyes, only with our hearts." Hilda told the mother to give Jesus the praise and to thank Him.

The young girl lived, and the story of the miraculous healing was told over and over in her village. Many others came to get medication at the Nazarene clinic. Each patient received a gospel tract that told about the living Jesus who could save them from their sins.

A young boy, Ramchandra, fell from a mango tree, breaking both of his wrists. The village doctor put some black tar and a cast on each arm and warned the lad's father not to remove the casts for several weeks. Due to extreme pain, the boy cried night and day; yet, the father was afraid to tamper with the casts. He became desperate and unsure how to help his son.

Someone told Ramchandra's father about Reynolds Memorial Hospital (RMH), and he hurried there with the boy. When Dr. Orpha Speicher removed the casts, the boy's skin was severely blistered, and both wrists were infected and swollen. Treatment began immediately and continued for several days.

> **Frequently, Hilda's day didn't end with a 12-hour shift, as medical emergencies were not confined to daytime hours.**

When the swelling went down, the doctor was able to see that Ramchandra's wrists and fingers were twisted out of shape. The lad could not even hold a pencil. Immediately, Dr. Speicher sent Ramchandra to Vellore, a treatment center for leprosy patients, where, in time, the use of the boy's hands was restored. Follow-up treatments were scheduled at RMH. When Ramchandra's hands were completely healed, he could write and draw again. Through this experience, Ramchandra accepted Jesus Christ as his personal Savior. He attended a Christian school and grew up to be an exemplary Christian.

Joy in the Daily Grind

A typical day for Hilda is difficult to describe, as each one was different and changed according to assignments. Her days certainly involved long hours. Often she reported for duty at 7 A.M., attended chapel, instructed nursing classes, worked in the operating room, administered anesthetics, and finished her work at 7 P.M. Frequently, Hilda's day didn't end with a 12-hour shift, as medical emergencies were not confined to daytime hours.

Other days involved teaching nurses on the wards, assisting with the outpatient department, checking workers' assignments, and inspecting the septic tanks to ensure they were working properly. Every day was filled with both major and minor problems.

When Hilda traveled to the outlying clinics, the day started at five in the morning. As many as 200 patients would be examined and treated no matter how long it took. She was careful not to turn anyone away. Fortunately, Hilda usually had a

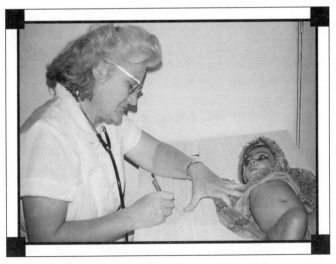

Nurse Moen working at an outlying clinic

driver to accompany her as she carried out this arduous and taxing routine.

In the middle of Hilda's hectic schedule, people often came to her home for tea. She would talk to them about the Lord and their needs. Opportunities for witnessing abounded. The days proved to be productive and rewarding for her. And even in the midst of the daily grind and mundane tasks, Hilda remained confident—and joyful—she was in the center of God's will.

Joy in the Midst of Disappointment

Hilda had a mild heart attack and experienced severe neck pain during the later years of her missionary career. To relieve the suffering, she was put in traction for many days. After weeks of great discomfort, she recovered. Later, she was giving anesthesia to a surgical patient. Hilda sat down, only to discover the expected stool was missing. Landing on the cement floor, she suffered a serious back injury. Hilda returned to Canada for an operation.

When Hilda furloughed in 1981, she learned she would never return to India. Mission officials decided to close many of the clinics, and this meant she was not needed on this particular field. In the same year Hilda was ordained at the Canada West District Assembly with the goal of returning to India as a preacher. This plan did not come to fruition either; therefore, she decided to take a leave of absence.

Preaching and pulpit supply in Nazarene churches in northern Saskatchewan became her new focus. In 1982 she was asked to be the supply pastor at the Saskatoon church. Two services each Sunday, Wednesday evening prayer meeting, and visitation consumed her time. The congregation was delightfully encouraging and responsive.

The same year Hilda found a part-time position as a night nurse at a health-care center in Saskatoon. These two nights of employment each week helped supplement her meager income. Later, at the same nursing home, she accepted a full-time, night-duty position, which she kept for six years.

Throughout this time, she never used traction for her chronic neck problem. When some lockers at work fell on her, her back was injured again. She was no longer able to be employed. Hilda took a disability retirement until she could receive a pension from the Canadian government. At the age of 65, she also received retirement from the Church of the Nazarene for 26 years of missionary service in India. Hilda now lives in Saskatoon close to her sister, Mina.

God has truly blessed the ministry of Hilda Moen. She feels the two and a half decades in India were the richest years of her life. "No, those years were not rich materially," she states. "Rather, they were rich because I believe I was doing exactly what the Lord called me to. Nothing else mattered. It did not matter if there were scary snakes, dreaded rats, or mysterious diseases. I experienced a rich and full life because I was in the center of God's will, satisfied and blessed in knowing I had helped people find Jesus Christ as their personal Savior. What a joyful journey I've had!"

A Journey of Abundance
Lenora Pease, India

"Dead? Our Laurie drowned? David too?"

Lenora spoke in incredulous, whispered tones, her voice filled with shock. "I can't believe it! The boys were so . . . *so* young! I just can't believe it!"

During the next few hours, Lenora and Bill Pease, numb with grief and despair, made prepara-

Lenora Pease

tions and buried their 15-year-old son, Laurie. They mourned with their good friends and fellow missionaries, Bronell and Paula Greer, as they also laid their son, David, to rest in India's soil. Together, the two couples journeyed through death's dark valley, not fully comprehending the tragic event that snatched their oldest sons from them. Still, they placed their faith in the sovereign God who had called and led them to the land of the Taj Mahal and Ganges River.

Just a few weeks before, after returning from a year's furlough, Bill and Lenora and their two sons, Laurie and Ken, vacationed in Bombay—a fun-filled holiday together as a family. Arriving home, the Greers asked if Laurie could go with them to evangelize in the villages, something the teenager really enjoyed. He liked to travel to the surrounding communities, especially with his good friend, David Greer, who was just a year older. Because it was just before Christmas, Bill and Lenora gave their consent—reluctantly.

On December 15, Laurie and David took their BB guns to shoot birds with the intent of having curried dove for supper. Near a lake, the teens spotted some doves that had flown over the water and settled on tall reeds. David thought he could get the doves, so he swam out toward them, thinking the reeds were on a small piece of land in shallow water where he could rest before swimming back. Since he was not a strong swimmer, he quickly got into trouble. Panicking and thrashing about, he became entangled in the reeds. According to eyewitnesses on the shore, Laurie tried for an hour to save his friend. He, too, became tired and sank down into the gray, murky water. Later, he was found at the bottom of the lake, his body in the position of treading water with reeds wrapped around his toes.

Once the two fathers arrived at the lake, they had to build a boat to go out and recover their sons' bodies. While they did this, the mothers went to Buldana 14 miles away to choose a burial plot and find some Christian men to dig the graves.

Bronell Greer was in such shock that he wanted to bury David right there on the lake's shore. With Bill's advice, the grieving fathers took both boys back to the Peases' home.

Laurie Pease, age 15

Bill arranged to have two coffins made out of the wood from a furlough packing crate. They sent a runner to summons a carpenter, and in a short time the craftsman made two modest caskets. Some Christian women lined the coffins with soft material. Since Indian law requires that a body be disposed of by sundown on the day an individual dies, a missionary doctor helped prepare the bodies for burial. The funeral was held at the Chikhli church. Friends and all the children and staff from the school plus many Hindus and Muslims from the village packed the sanctuary.

Before this catastrophic event, Lenora and her colleagues had longed and prayed for revival. Some of the boys at the mission school had been misbehaving. Also, some of the nurses were having personal problems. The missionaries met together, interceding specifically for revival. "Dear Lord, at any cost, give us a revival. Yes, at any cost," Lenora remembers praying.

Laurie and David's deaths were not in vain.

On the Sunday following Lenora's prayer, the Peases attended the Maiaker church in a nearby village. As Lenora spoke during the morning service about heaven, God's presence filled the sanctuary. The next Saturday the boys drowned.

Since the Pease and Greer families were well known, news spread swiftly that the two missionary kids had died. In the Indian culture, the loss of the oldest child and particularly a boy was of great importance. What a trauma to everyone!

Revival came to India—at the school, the hospital, the mission station, and even the boarding school for missionary kids. People throughout that region of India were saved. God had answered prayer. Laurie and David's deaths were not in vain. The Peases and Greers experienced God's abundant grace and peace.

To Lenora, what mattered most was the fact that Laurie knew Jesus Christ as his personal Savior. He had made his peace with God the summer of 1962 while on furlough at Harmattan Nazarene

The Pease family with Dr. George Coulter

Youth Camp just a few months before his untimely death.

Bill Pease with sons Ken, 17, and Rick, age 4 months

Two years after Laurie died, Lenora and Bill had another son, Richard Lynn. (Since he was born in India, he has dual citizenship.) Ken, then age 17, welcomed this brand-new brother. Lenora thought of this latest baby as a miracle because of her age and her many struggles to have a third child. Although no child could take the place of Laurie, the birth of Rick helped to fill the void in the Pease home and hearts.

What brought Lenora and Bill to India? Let's go back in time five decades.

The Journey into This World

A blinding blizzard, ferocious yet common, struck the Canadian prairies. Snowdrifts stacked above fence posts and piled up across highways and country roads. No one would be traveling this night. Amid the storm, Emma had her baby, unexpectedly early. No doctor was able to traverse the seven miles from town to the farm on time, so Emma's mother delivered the child, cleaned her up, and placed her in her mother's arms.

The newborn was wrapped in a cotton cloth and placed in her grandfather's shoe box, which became her first bed. Her head was so tiny it easily fit in the palm of one's hand or inside a teacup. No incubator was available, so they kept the tiny baby warm by placing her on the oven door of an old iron stove in the farmhouse kitchen.

"Emma, if this little girl lives," Grandmother remarked, looking down at this premature infant, "we'll give her to God to be a missionary." This was the beginning of the life of Lenora Kaechele Pease, born in Bashaw, Alberta, on February 28, 1919.

The Growing Up Journey

Lenora was the second child in Solomon Kaechele's family. Her father, a minister in the Church of the Nazarene, was one of the pioneer preachers in Alberta at the time. When her dad, a strong Holiness preacher, took Lenora with him on pastoral visits, they traveled by horse and buggy in the summer and cutter in the winter. Lenora watched as he prayed and dealt with the heartaches of his people. Constantly rubbing shoulders with the missionaries, preachers, and evangelists who came into the parsonage made a significant impression on the young girl. She was already in training for the work God planned for her.

As a child, Lenora recalls living in primitive houses where they engaged in combat to eliminate bugs and other pests. One home was a converted barn. Although she never had much as a young girl, she never felt deprived and learned early on to never beg nor ask for anything.

One time, the Kaechele family didn't have anything to eat, and Lenora was hungry. *How strange,* she thought, *although Dad is serving God, still there's no food.* At suppertime, her mother put on the tablecloth and set the table, placing a beautiful bouquet of flowers in the center. Her father called all of the family to come and sit down, even though the table lacked anything edible. They bowed their heads and thanked God for what they were going to eat. Lenora remembers peeping to see if food had appeared, but, alas, nothing. While her dad prayed, there was a knock on the door. When Rev. Kaechele opened it, he found a large hamper of food on the porch. Peculiarly, no one was there, and the family never learned who brought the "heavenly manna." God answered the prayer and honored the faith of the godly parents. As a result, the children learned a valuable lesson about trusting God—a marvelous example of His abundant provision.

After Lenora completed grade 10, her mother became quite ill. Being the oldest girl, Lenora stayed home and cared for the family. When time came for her to resume her education, her dad did not have the finances. "I have an old, white cow that I can sell," he told her. "You can use the money for the down payment on your education. It isn't very much, but it's a start."

What fun she had arranging dates for all the suitors!

Off Lenora went to Canadian Nazarene College (CNC), in Red Deer, Alberta. Eight years later she completed both high school and college. To earn the extra money she needed, she clerked in a store, cleaned houses, and ironed white shirts.

A Journey of Romance

While in college, Lenora was the assistant dean of women. The "dating system" worked like this: if a young fellow wanted to date someone, by Thursday evening he had to ask the dean of women and arrange the date through her. One weekend when the women's dean, Agnes Comfort, was away, Lenora was left in charge. What fun she had arranging dates for all the suitors!

"Miss Kaechele," someone yelled up the stairs. "Bill Pease is out here and wants to see you."

Lenora wondered who the young woman was. She went tripping down the stairs. "Well, Brother Pease, whom would you like to take out this weekend?" she asked.

"You," he announced.

Not having a clue of Bill's interest or intent, Lenora nearly fainted on the spot. "I . . . I need some time to . . . to think," she responded.

Bill and Lenora had a date that Friday night, and soon they became a couple. Their romance developed into a lasting love. Lenora, age 27, married Bill Pease on July 3, 1946. Dr. William Noble King performed the ceremony, and some of their good friends participated. Both Fred Parker and Alice Johnson sang, and Jean Parker played the piano. The reception was held at CNC.

Soon after their wedding, Bill and Lenora went to Dawson Creek, British Columbia, as pastors. Two years later, in 1948, they went to Drumheller, Alberta. They also pastored in Claresholm, Alberta, from 1949 to 1954.

The Journey to India

"During an evening service at the family camp in Red Deer," Lenora recalls, "the Spirit urged me to go forward. As I prayed at the altar, God spoke, and I responded with a definite yes. That settled my call to be a missionary."

Bill and Lenora loved being pastors of local churches in Canada. Settling down there could have been easy. But God had other plans for this couple, and His call to missions never dimmed.

When they were asked to meet with mission officials of the Church of the Nazarene in Kansas City, both of their boys, Laurie and Ken, had the measles. Bill went by himself for the interview. Later, he telephoned his wife. "Lenora, they want us to go to India."

"India, Bill? . . . India? Where . . . where is that?" she questioned.

All along Bill and Lenora had prepared their minds and hearts for Africa. Bill detected misgivings in the voice of his helpmate. "I think we should pray about it," he said. "But we don't have long, Lenora. The mission board wants our answer in 24 hours."

"I shut myself away," Lenora remembers. "I got someone to look after the boys while I prayed and fasted. I continued in prayer until a peace—God's abundant peace—swept over me. When Bill called the next day, I was able to tell him, 'I'm ready to go to India.'"

The day Lenora put her feet on India's soil, she

loved it and adjusted quickly to this new and exotic country. She and Bill were an older couple, having waited 10 years before being assigned. Their personalities blended beautifully with the Indian people. Working well together, Bill and Lenora complemented each other's gifts and strengths.

The Journey of Missionary Service

Bill's first assignment in India was handling the treasurer's books. His years as treasurer of the Canada West District working with Dr. Edward Lawlor provided him invaluable experience in business administration. Later, Bill was assigned to a short-term position at a hostel (dormitory) for boys while the youth attended high school and college.

Another assignment was the 10-year span Bill and Lenora Pease served at a Nazarene educational

Lenora Pease with Indira Gandhi, India's prime minister

institution where there were 350 boys and girls. Bill was the principal, and Lenora taught English with the main responsibility for the girls' hostel. Again she was involved with a boarding school similar to her years at CNC, but only this time with children.

Without question, prayer is a vital part of a missionary's life. Many times Lenora found herself and her family in situations where she depended solely on prayer. She also knew that people at home praying was essential. Often, in her spirit, she sensed that family and friends were interceding for the Pease family. They would tell her that on a certain day God awakened them and prompted them to pray for her, Bill, and the boys.

On one occasion when Bill and Lenora were working at the school, a loud noise in the middle of the night came from the boys' hostel. Some of the fellows rushed over to the Pease home, screaming at the top of their lungs, "Sahib! Sahib! Come right away!"

Bill bounced out of bed, pulled on his pants, slipped into his sandals, and sprinted out of the house. He yelled back to Lenora to bring the flashlight. As he dashed down the stairs, he stepped on something, but without stopping hurried on to the boys' dormitory.

Unaware of what he had stepped on, Bill and Lenora alerted the others to the possibility that it had been a snake. Four days later, they found a king cobra in the hedge in front of their house. They killed it and stretched it out on the ground. Looking closely at this impressive creature, they noticed a footprint right behind its head—the imprint of Bill's sandal.

❦

> **Immediately, God touched her, her fever left, the chills stopped.**

❦

If he had stepped on the snake at any other place than the hood, it could have easily struck him, and death would probably have been certain and rapid. "I know someone was praying for Bill's protection that day," Lenora says, "and I believe I will discover who prayed when I get to heaven."

On another occasion, Lenora was writing her preliminary language exam for Marathi. In addition to the pressure of the exam, she had malaria with a temperature of 104 degrees. She asked God to burden someone at home to pray for her. Immediately, God touched her, her fever left, the chills stopped. She wrote her exam—*and passed!* About a month later, Lenora and Bill received a letter saying that on this certain day, at this certain time, this individual was praying for them. Once again, God answered prayer—abundantly.

The Retirement Journey

Lenora never gave much thought to retirement. In her mind, she always thought she and Bill would retire together. They hoped that when the time came to join Laurie in heaven, God would take both of them at the same moment. They just looked forward to retiring together and doing all the things they never had time for.

In 1979 doctors discovered a large tumor in Bill's liver. The physician suggested that the Peases go home immediately, rather than hospitalizing Bill in Bombay. Lenora rushed to make arrangements to fly to Canada. Nearly losing her husband in Frankfurt, Germany, Lenora called their son Ken and told him to arrange for an ambulance to meet the plane, explaining that his father was very ill.

Not fully realizing how sick his dad was, Ken met the plane by himself. By this time, Bill's condition had worsened. Ken picked up his dad, placed him in his van, and rushed to the hospital. Bill died four days later.

Lenora, though grieving, went right on with the full deputation schedule she and her life's partner had committed to. When the furlough was over, she returned to India two years later to bring closure to that part of her life.

Today, Lenora resides in Calgary. She still keeps busy, even though *retirement* is not her favorite term. She prefers to think that she is just *retreaded.* Active in her local church, she also travels, speaking in churches.

"I always focused my life on obedience to the call of God," Lenora testifies. "There were many hard times. But I was confident that God would be with me through it all, that His abundant grace was mine. A Scripture promise when I first went to India is one I still cling to: 'So do not fear, for I am with you; do not be dismayed, for I am your God. I will strengthen you and help you; I will uphold you with my righteous right hand'" (Isaiah 41:10).

Lenora Pease's life—truly a journey of abundance.

A Never-Ending Journey

Mary Wallace, Nicaragua and Chile

Frustration. Bewilderment. Fear. Anxiety. Anger.

Mary Wallace encountered a range of emotions as she struggled to understand the difficult dilemma she faced. *Why all the harassment?* she wondered. *Why are people trying to run me out of town?*

Mary Wallace

This was one of the most distressing and troubling times in Mary's 33 years as a Nazarene missionary. The Christian school that she supervised in a small city, well known for its discipline, enjoyed a good reputation. Yet, some of the local residents decided Mary should leave. Her enemies planned extreme and vexing strategies, and the annoyances began. For example, they arranged to have pamphlets dropped from airplanes in an attempt to influence parents to disenroll their children. Others canvassed the students' homes and offered them incentives, such as shoes and books, to change schools. They knew if the school closed, Mary would be forced to move.

Part of the harassment included irritating nuisances and even break-ins directed at Mary's home. Even though she stationed a guard at her patio door, the troublemakers were never caught.

A friend insisted on loaning her a gun for protection. He took pictures of her with the firearm in front of her home. He wanted everyone to know Mary had a weapon. Little did the townspeople know that Mary was more afraid of the gun than she was of the prowlers.

Mary marched to the police station and announced that she had a gun. In her frustration, she simply took the steps she felt were needed to plant fear and respect in the people. She was determined to stop the harassment so the work of the school could continue.

Night after night, Mary's sleep was disrupted. She knew her teachers had contracts and expected their salary. Yet, without students to pay fees, no funds would be available to operate the school.

One night, God spoke distinctly to Mary. "Daughter, whose work is this?"

"Yours, Lord."

Reading her Bible, Mary came across Psalm 27:13-14, where David provided a word of encouragement: "Be strong and take heart and wait for the LORD." She laid her Bible down and breathed a prayer of thanksgiving, "Thank You, God. It's in Your hands." Mary fell fast asleep. The prowling stopped, and next year the school had record enrollment.

A Lifelong Journey

For 35 years, Mary Wallace fully dedicated her life to Nazarene mission service—25 years in Nicaragua and 10 years in Chile. Her missionary journey took her across tall mountains and turbulent rivers, through dense jungles and isolated deserts. Mary traveled by horseback, jeep, boat, and

Mary's love for the people was as big as the God she served.

foot. The destinations included tiny villages where she started schools or teeming cities where she sold literature. The fruit of her labor can be measured by the myriad graduates who are now church and business leaders and by the countless Holiness books that have been purchased and perused.

Over the years Mary's assignments varied greatly—developing curriculum, maintaining financial records, instructing graduate students, teaching music, training teachers, conducting retail business, and planting churches. It seemed as though Mary became one with the people as she became highly proficient in their mother tongue, communicating freely in the people's heart language. For Mary, going to work each day was the joy of her life. Her love for the people was as big as the God she served. Indeed, hers was a never-ending journey of faith and adventure.

Mary going to visit a rural school

The Journey Begins

Florence Mary Loreen Wallace, best known as Mary, was born in Belmont, Ontario, on February 8, 1924. Her parents, Hugh Morrison and Jean Harkness Wallace, had two other girls: Isabel and Winnifred. When Mary, the middle child, was nine years old, her mother passed away. With the help of housekeepers, her father raised his three daughters.

Throughout most of her childhood, the Wallaces attended the local Presbyterian church. One night when a missionary from Formosa (now Taiwan) spoke at the church, Mary decided she would be a missionary nurse. This was her earliest impression that God wanted her to do something very special.

At the age of 15, the Wallaces moved just outside the city of London, Ontario. A neighbor introduced them to the Church of the Nazarene and its pastor, Rev. Ernest Collins. Soon after, Mary became a Christian, asking Jesus to come into her life. She opened her heart to God, and He continued to transform her, preparing her for the work to which He would call her.

After graduating from grade 12 in 1941, Mary took a commercial course that resulted in a position with a local law firm. A year later, she took a job with an insurance company, where she worked for seven years.

During this time, Joyce Blair, a Canadian missionary in Belize, came to Mary's church. As Joyce spoke in the service, Mary received a clear message to teach children in Central America. God continued to prepare Mary for the specific task He had planned for her.

In January 1949, Mary entered Eastern Nazarene College to prepare for teaching. Upon graduation Mary accepted a position with the Free Methodist Church in a secondary boarding school, where she was responsible for 20 girls in a dormitory. This two-year assignment was strategic in fulfilling the experience requirement before becoming a missionary in the Church of the Nazarene.

Mary encountered a setback in February 1954 when she received word that single women were not being sent as missionaries and there was no need for teachers at the time. Discouraged and disappointed, Mary thought that if she waited too long, she would be too old to be assigned. Calculating her next step, she answered an advertisement in the Toronto newspaper for a teacher in a gold-mining community in the jungles of Central America. As the successful applicant, Mary moved to Nicaragua to teach grades six through nine.

Excitement began to build for Mary. In Nicaragua, she was able to contact the Moravian missionaries who worked in the villages and communicate with Nazarene missionaries in the western part of the country. Several times a year she visited Nazarenes for fellowship and to make the critical contact that would eventually lead to work as a Nazarene-appointed missionary. Finally, when the Nazarene mission school needed a teacher, the missionaries notified Mary. She promptly applied, and Miss Wallace was hired on a temporary basis.

With her proverbial foot in the door, this new assignment put Mary in the place to consider other positions. When Esther Crain left the mission field, Mary took over the supervision of the Nazarene school in the city of Rivas. In June 1955, Mary officially met the mission board and was assigned to Nicaragua. She spent the first two months in Mexico in intense language study. While it seemed Mary came through the back door, she was finally on the mission field for her beloved Church of the Nazarene. Her journey now took on new dimensions.

The Journey with Carmen

Seven-year-old Carmen started attending the backyard Sunday School in Rivas that Mary directed. The young girl brought her little brother, often nestled on her hip, and many of her neighborhood friends. Faithful in attendance, Carmen attentively listened to every lesson.

One Sunday when Carmen did not show up, Mary visited the girl's home, a one-room dwelling with a cooking area. The missionary learned that Carmen had been severely burned earlier that week. A pot of boiling beans accidentally toppled off the stove, spilling over Carmen's arm. Her father picked her up and whisked her outside; an ambulance came and took her to the hospital.

> "I don't pray that way," Carmen insisted. "I pray in Jesus' name."

Carmen's mother sat by the bed until her daughter regained consciousness. When Carmen awoke, she whispered, "Dear Jesus, help me." She continued to pray as she suffered with excruciating pain. Over time she gradually improved, and the burns began to heal. Each day the Catholic sisters in charge at the hospital visited the girl and tried to teach her to pray by crossing herself.

"I don't pray that way," Carmen insisted. "I pray in Jesus' name."

The sisters continued their daily visits, encouraging Carmen to pray *their* way, but she was not interested. The sisters, being concerned, asked the priest to call on her.

Carmen informed the priest that she prayed in Jesus' name. But the priest persisted and continued to come every day. His visits troubled Carmen. Taking matters into her own hands, she reacted by pulling the bedsheet over her head and pretending to be sound asleep when the priest entered the ward. Eventually, a sister said to Carmen's mother, "You must be wonderful parents to have taught your little girl such faith."

The mother began to cry. "No, we aren't. We have *not* been good parents."

"Where did your little girl learn to pray as she does?" the sister asked.

"Oh, she attends the Nazarene Sunday School," the mother replied, "and the lady there has taught her to pray."

With her scarred arm, Carmen went to visit Mary. The missionary nurse carefully treated it with some soothing salve to help keep the arm mobile, and she sent some of the ointment home so Carmen's mother could provide ongoing care. The arm healed well with no apparent loss of use.

During the next year, Carmen suddenly and mysteriously was absent from the backyard Sunday School. Mary was disappointed to learn that the family had moved.

Many years later at a conference Mary attended, a fine-looking gentleman, dressed in a white shirt and tie, approached and called her by name. Mary did not recognize him. When she inquired about his identity, he smiled and stated that he was Carmen's father.

Masses of people, dazed and confused, rushed about aimlessly and screaming in despair. Chaos reigned.

The man reported several happenings, all of them great news to Mary. First, he said that Carmen was able to use her arm to do many tasks, including the family laundry. Second, he related that, because of Carmen's testimony in the hospital, he and his wife began to attend church services. They had made the decision to become Christians, and he had been delivered from alcoholism. Third, he informed the missionary that he was now a lay pastor, studying to become a full-time minister. Mary rejoiced and praised God, for He had used little Carmen in miraculous ways.

A Journey Through an Earthquake

The earth trembled and shook. Buildings tottered and then toppled. Streets and roads split in half. Fires broke out everywhere. The city became a blazing inferno. Masses of people, dazed and confused, rushed about aimlessly and screaming in despair. Chaos reigned.

Managua, the capital of Nicaragua, was under siege from the devastating earthquake that shook the world on December 22, 1972. The damage, extensive and mind-boggling, included total destruction of 600 blocks of the city's core and partial ruin of another 600 blocks in the surrounding area. People searched diligently for loved ones. The highways in and around Managua were jammed with vehicles leaving and entering the city.

Mary had transferred to Managua four years earlier to start a Christian bookstore. The bookshop and Mary's home, located above the shop, were located in the center of the quake. Providentially, she escaped without injury, even though buildings collapsed all around her. She was overwhelmed by the cries of the injured roaming the streets and people still trapped inside the rubble.

After many hours and numerous aftershocks, Mary crawled back into the store and clambered up to the second floor despite the fact that the nondisappearing stairway had now disappeared. She collected clothing and necessities for her friends from Costa Rica who were visiting for Christmas, but in her distress she picked up two white shoes, both for

Dedication of the new bookstore after the earthquake

the same foot, and completely forgot any garments for herself. Mary stayed at the Nazarene Bible School for several weeks, all the while thinking about the time she would be able to reopen the bookstore.

When the roads were repaired, Mary drove to San José, Costa Rica, where book depositories and supply stores were located. She picked up a carload of books to take back to Managua. With all the bookstore furniture destroyed, she devised a makeshift table of two sawhorses, a board, and a piece of oilcloth. She was back in business.

A man entered the store, folded his arms across his chest, and looked down at the sawhorses. "Now, that's what I like," he said.

"You like the sawhorses?" Mary asked.

"No, getting started is what I like."

Mary's face brightened with an enormous grin. "Me too," she responded.

While at Managua, Mary organized the seventh Nazarene church. She remained in this city until June 1979, when the Canadian government ordered her to leave because of the escalating civil war. Though disappointed at leaving, Mary was pleased that the bookstore ministry would continue, because it had grown to become a self-supporting store with seven full-time workers.

The Journey to Chile

The World Mission Division gave Mary a new assignment in the southern hemisphere at the Nazarene seminary in Santiago, Chile. Though she missed the Nicaraguan people very much, this opportunity for training pastors was most rewarding. She taught at the seminary from 1980 to 1985.

Wherever Mary's journey took her, she always seemed to be involved with bookkeeping and treasurer's jobs. In Chile she established separate accounting systems for the district and the mission. She also competently completed all necessary forms and financial reports for the government so the Church of the Nazarene could maintain its legal status in the country.

One of Mary's biggest challenges dealt with the sale of a piece of property to the government for use as a traffic circle that involved several million pesos. With the advice and assistance of a friend in a local bank, Mary set up investment programs that facilitated the future financing of prop-

Mary with some of her students in Chile

erties and construction projects for the Church of the Nazarene in Chile.

For one year, 1986, Mary traveled and taught theological extension courses in different parts of the country.

The year 1987 brought a new assignment: starting the work of the Church of the Nazarene in Viña del Mar, Chile. Mary pioneered the work as there were no Nazarenes, no money, no property. After five months, God led Mary to a home where she began Bible studies. In the next three months, 28 people had accepted Christ. Mary and the people started Sunday morning services and a Sunday School. Within one year, the church was organized, and one of the new converts Mary trained was the new pastor.

The city of Valparaíso was nearby, and Mary worked with the Nazarenes there as well. She provided encouragement to the young pastor and the members. When the church received some Alabaster money, the people purchased a lot and put up a tent. A Work and Witness team came and helped build a two-story structure, a chapel on the first floor and parsonage on the second.

The Journey Ends? Never!

Mary returned to Canada in early 1989 for a much-needed furlough. She spent the year traveling and speaking to congregations in Canada and the United States. On February 1, 1990, Mary retired after 35 years with the Church of the Nazarene. She has spent the last several years working in her local church in London, Ontario, and with immigrants who speak Spanish.

Mary's involvement with the Nicaraguan people did not cease with her retirement. In February 1995, she returned with a Work and Witness team to restore a school building that had been vandalized during the civil war. The school she had helped to start was still running with just over 200 children enrolled. On a subsequent trip the same year, Mary was honored during graduation ceremonies and at a banquet with over 200 in attendance. She was delighted with the possibility that this school would become a secondary school, complete with computer and technical courses.

Mary Wallace now lives in London, Ontario, where she lives a healthy, comfortable life. "I love my God, I love my work, and I love the Church of the Nazarene," she says. Ministering to others and sharing the love of Jesus Christ never changes no matter where Mary may be on her never-ending journey. ♣

Pronunciation Guide

The following information is provided to assist in pronouncing unfamiliar words in this book. The suggested pronunciations, though not always precise, are close approximations of the way the terms are pronounced in English.

Alymer	AL-mer
Anusayabai	AN-uh-SIE-uh-bie
Bansod, Eby	BUHN-sohrd EE-bee
Belize	buh-LEEZ
Benque Viejo del Carmen	BEHN-kay vee-AY-hoh dehl KAHR-mehn
Bhakshibai	BAHK-shee-bie
Buldana	bool-DAH-nuh
Chappell	shuh-PEHL
Chikhli	CHIHK-lee
Clandeboye	KLAN-dee-boy
Claresholm	KLEHRS-hohlm
Drumheller	DRUHM-HEHL-er
Endzingeni	EHN-zihn-GEH-nee
Friesen	FREE-zuhn
Ganges	GAN-jeez
Harmattan	hahr-MAT-uhn
Kaechele	KAYK-lee
Kashewala	kah-sheh-WAH-lah
Lacombe	luh-KOHM
Madras	muh-DRAHS
Maiaker	MAY-uh-ker
Malkapur	MAL-kuh-poor
Managua	muh-NAH-gwuh
Manzini	mahn-ZEE-nee

Marathi	muh-RAH-tee
Mina	MEE-nuh
Moen	MOH-ehn
Nanaimo	nuh-NIE-moh
Oliphant	AH-luh-fuhnt
Pease, Lenora	PEEZ luh-NOHR-uh
Pundit	PUHN-diht
Pusad	POO-suhd
Ramchandra	RUHM-CHUHN-druh
Rivas	REE-vahs
Sahib	sah-HEEB
San José	SAN hoh-SAY
Santiago	SAHN-tee-AH-goh
Saskatchewan	suh-SKACH-uh-wahn
Saskatoon	sas-kuh-TOON
Taj Mahal	TAHZH muh-HAHL
Valparaíso	VAHL-pah-rah-EE-soh
Vellore	vuh-LOHR
Vilakati	VIHL-uh-KAH-tee
Viña del Mar	VEE-nyah del MAHR
Washim	WAHSH-uhm
Witte	WIHT-ee